New Original Graphs in Trigonometry

Volume 1

Chibamba Mulenga, Ph.D.

New Original Graphs in Trigonometry

Volume 1

Author's logo Math Marvels

Ace Book Publishers
100 Church Street
8th floor
Manhattan, NY, 10007

www.acebookpublishers.com

PRINTED IN THE UNITED STATES OF AMERICA

DEDICATION

In loving memory of my mother, Mwelwa Mitalabi; my uncle Chewe Mitalabi; and my grandmother, Kafi Nsankinta, a storyteller par excellence of Bemba royal lineage… and to wayfarers in the ocean of life who rendered me kindness and support.

ACKNOWLEDGMENTS

I am grateful to the technical experts at Ace Book Publishers for making the publishing and marketing of this book possible.

TABLE OF CONTENTS

PREFACE

The purpose of this book is to introduce new original graphs in trigonometry. The new original graphs are related to graphs of the six basic trigonometric functions. These are the sine, cosine, tangent, cosecant, secant, and cotangent functions. Let x be a real number. Then sinx, cosx, tanx, cscx, secx, and cotx are trigonometric functions with known definitions and graphs.

This book excludes those familiar graphs. Instead, the book presents new graphs that I invented in trigonometry. I received copyrights for the graphs from the United States Library of Congress Copyright Office between the years 2011 and 2014.

The new original graphs in trigonometry occurred to me as follows. One evening, the inner voice prompted me to sketch the graph of sin|x|, where the expression |x| denotes the absolute value of x. Although I had taught trigonometry for several years, I had not heard of the function sin|x| and its graph. Nevertheless, I sketched the graph of sin|x| as I moved to do so.

I made two observations about the graph of sin|x|. The first is that, unlike the graph of sinx, the graph of sin|x| is symmetric about the y-axis in the x-y rectangular coordinate system. The second observation is that the related mathematical literature does not indicate any instance where the graph of a sine function is symmetric about the y-axis.

Based on the two observations, I deduced that the graph of sin|x| is a new graph in trigonometry. Consequently, the function sin|x| is a new trigonometric function. The conclusion follows from the one-to-one correspondence between a graph and a function from which that graph is generated. Accordingly, the function sin|x| and its graph are two new mathematical concepts in trigonometry.

Several days after sketching the graph of sin|x|, I sketched the graph of cos|x| out of curiosity. I observed that cos|x| and cosx have the same graph. Since they have the same, then cos|x| and cosx are the same function with two different algebraic expressions. As cosx is not a new function, then cos|x| is also not a new function. Likewise, sec|x| and secx have the same graph. Since they have the same graph, then sec|x| and secx are the same function with two different algebraic expressions. As secx is not a new function, then sec|x| is not a new function as well.

We now consider two trigonometric identities, which are alluded to in the preceding paragraph. The first identity is that cos|x| = cosx. This identity follows from the discovery by graphing that cos|x| and cosx have the same graph. The second identity is that sec|x| = secx. This identity follows from the discovery by graphing that sec|x| and secx have the same graph.

To verify the two identities, let x be a real number. Then by definition |x| = x if x ≥ 0, and |x| = -x if x < 0. Thus, cos|x| = cosx if x ≥ 0, and cos|x| = cos(-x) if x < 0. But from unit circle trigonometry, cos(-x) = cosx. Therefore, cos|x| = cos(-x) = cos(x) if x < 0. Since cos|x| = cosx if x ≥ 0, and cos|x| = cosx if x < 0, then cos|x| = cosx, for any real number x. To verify that sec|x| = secx, we use the result that cos|x| = cosx. By definition sec|x| = 1/cos|x|, for cos|x| ≠ 0; and sec|x| = 1/cosx, for cosx ≠ 0. Since cos|x| cosx, then sec|x| = 1/cos|x| = 1/cosx = secx, for cos|x| ≠ 0, and cosx ≠ 0, as desired.

Inspired by inventing the function sin|x| and its graph, I formulated other suitable trigonometric functions for the purpose of sketching their graphs. It's from the suitable trigonometric functions that I produced other new original graphs in trigonometry.

I said that the new original graphs in trigonometry are related to graphs of the six basic trigonometric functions. For that reason, several books that contain graphs of the six basic trigonometric functions are presented in this treatise as Related Books. However, the related mathematical literature does not refer to any of the new original graphs in trigonometry.

It's worthwhile to mention that I hand-sketched all graphs that I submitted to the United States Library of Congress Copyright Office for copyright consideration. To produce a satisfactory graph, I evaluated a function at as many points as needed from each quadrant of the unit circle in its standard position on the x-y coordinate axes.

However, the majority of graphs in the book were computer-generated, and the rest were hand-sketched. The computer-generated graphs were produced when I had access to suitable computer software. When that access expired, I hand-sketched the remaining graphs. Then I scanned and turned each hand-sketched graph into a Word document. Finally, I combined computer-generated graphs and Word document graphs into a book manuscript.

This book constitutes Volume 1 of New Original Graphs in Trigonometry. The book introduces new concepts, definitions, trigonometric identities, discoveries, and examples. The book is suitable for readers who have studied unit circle trigonometry. The book can serve as a new pedagogical resource in existing high school and college math curricula. Mathematicians can use the book as a starting reference point from which to extend its discoveries in different directions.

The appearance in this book of the new original graphs in trigonometry marks a new page in the history of trigonometry. Through its innovative concepts, definitions, discoveries, and examples, the book has great potential to inspire its readers to pursue careers in diverse fields such as mathematics, physics, astronomy, land surveying, weather patterns forecasting, and oceanography.

I thank Ace Book Publishers for allowing my logo, MATH MARVELS, to appear on the back cover of the book above the barcode. The logo is my marketing idea for attracting attention to the book. I hope that readers will enjoy this logo and endear it to their hearts.

The given original graphs, of a completely new and unique type in the mathematical firmament, will echo, surprise, confound, and inspire the worthy inquirer throughout the irreversible cycles of time.

To the reader who is eager to know more about these amazing and extraordinary graphs, we present Volume 1 of New Original Graphs in Trigonometry.

Chibamba Mulenga

CHAPTER ONE

GRAPHS OF SIN|X|, CSC|X|, TAN|X|, AND COT|X|

1. sin|x|

Definition: Let x be a real number. Then the function h defined by

$$h(x) = \sin|x|$$

is called the sine of the absolute value function.

The domain of h is $(-\infty, \infty)$, and its range is $[-1, 1]$. The graph of $y = \sin|x|$ is symmetric about the y-axis.

EXAMPLE: $h(0) = 0$, $h(\pi/2) = 1$, $h(\pi) = 0$, $h(3\pi/2) = -1$, $h(2\pi) = 0$, $h(-\pi/2) = 1$, $h(-\pi) = 0$, $h(-3\pi/2) = -1$, and $h(-2\pi) = 0$. The graph of $y = h(x)$ is shown in Figure 1.1:

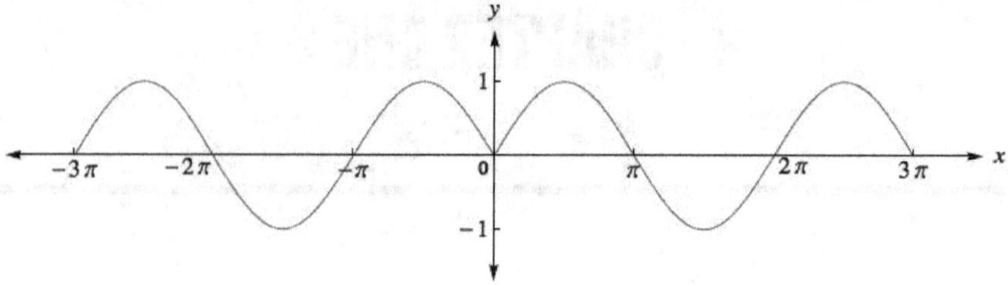

2. csc | x |

Definition: Let x be a real number. Then the function h defined by

$$h(x) = \csc|x|,\ \csc|x| = 1/\sin|x|, \text{ and } \sin|x| \neq 0$$

is called the cosecant of the absolute value function.

The domain of h is {all real x: $x \neq \pm n\pi$, n = 0, 1, 2 . . .}, and the range of h is

$(-\infty, -1] \cup [1, \infty)$. The vertical asymptotes of h are at $x = \pm n\pi$, n = 0, 1, 2 . . .

EXAMPLE: $h(\pi/6) = 2$, $h(\pi/2) = 1$, $h(5\pi/6) = 2$, $h(7\pi/6) = -2$, $h(-\pi/6) = 2$, $h(-\pi/2) = 1$, $h(-5\pi/6) = 2$, and $h(-7\pi/6) = -2$. The graphs of y = h(x) and y = sin | x | are shown in Figure 1.2:

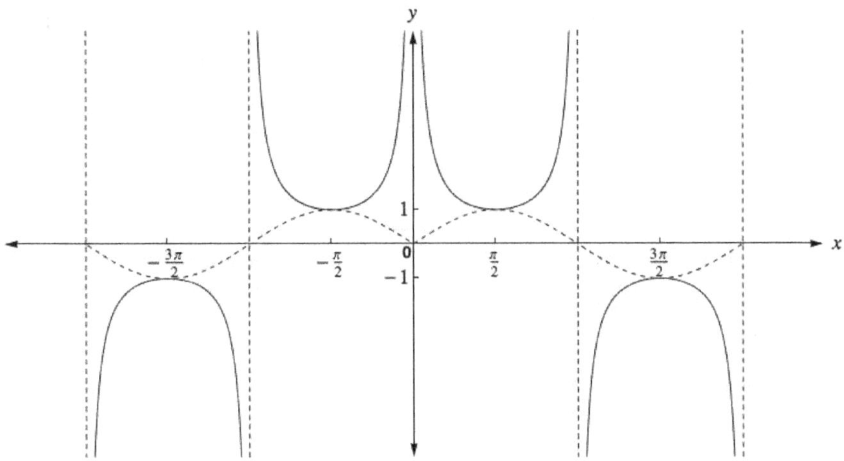

3. tan | x |

Definition: Let x be a real number. Then the function h defined by

$$h(x) = \tan|x|$$

is called the tangent of the absolute value function.

The domain of h is {all real x: $x \neq \frac{\pi}{2} \pm n\pi$, n = 0, 1, 2 . . .}, and the range of h is $(-\infty, \infty)$. The vertical asymptotes of h are at $x = \frac{\pi}{2} \pm n\pi$, n = 0, 1, 2 . . .

EXAMPLE: h(0) = 0, h(π/4) = 1, h(3π/4) = -1, h(π) = 0, h(-π/4) = 1, h(-3π/4) = -1, and h(-π) = 0. The graph of y = h(x) is shown in Figure 1.3:

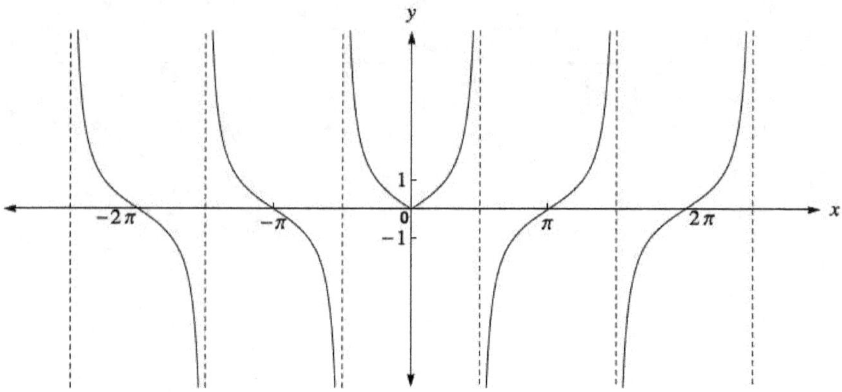

4. cot|x|

Definition: Let x be a real number. Then the function h defined by

$$h(x) = \cot|x|, \text{ and } \cot|x| = \tan(\tfrac{\pi}{2} - |x|)$$

is called the cotangent of the absolute value function.

The domain of h is {all real x: x ≠ ± nπ, n = 0, 1, 2 . . .}, and the range of h is (-∞, ∞). The vertical asymptotes of h are at x = ± nπ, n = 0, 1, 2 . . .

EXAMPLE: $h(\pi/4) = 1$, $h(\pi/2) = 0$, $h(3\pi/4) = -1$, $h(5\pi/4) = 1$, $h(-\pi/4) = 1$, $h(-\pi/2) = 0$, $h(-3\pi/4) = -1$, and $h(-5\pi/4) = 1$. The graph of $y = h(x)$ is shown in Figure 1.4:

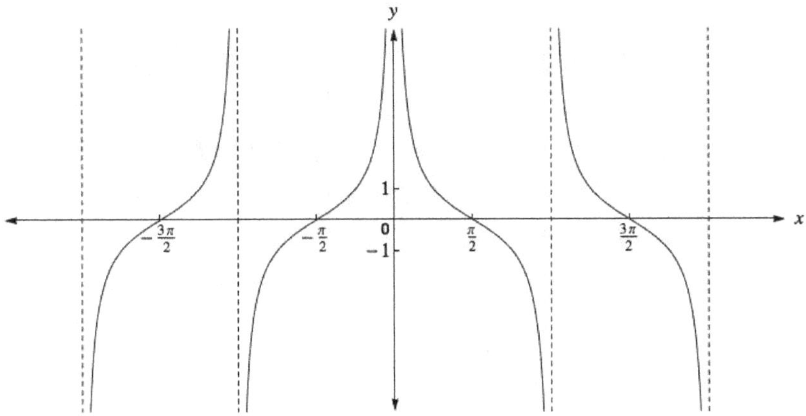

CHAPTER TWO

GRAPHS OF | SINX | , | CSCX | , | COSX | , | SECX | , | TANX | , AND | COTX |

1. | sinx |

Definition: Let x be a real number. Then the function h defined by

$$h(x) = |\sin x|$$

is called the absolute value of the sine function.

The domain of h is $(-\infty, \infty)$, and its range is $[0, 1]$.

EXAMPLE: Let h(x) = h(0) = 0, h(π/6) = 1/2, h(π/2) = 1, h(5π/6) = 1/2, h(π) = 0, h(-π/6) = 1/2, h(-π/2) = 1, h(-5π/6) = 1/2, and h(-π) = 0. The graph of y = h(x) is shown in Figure 2.1:

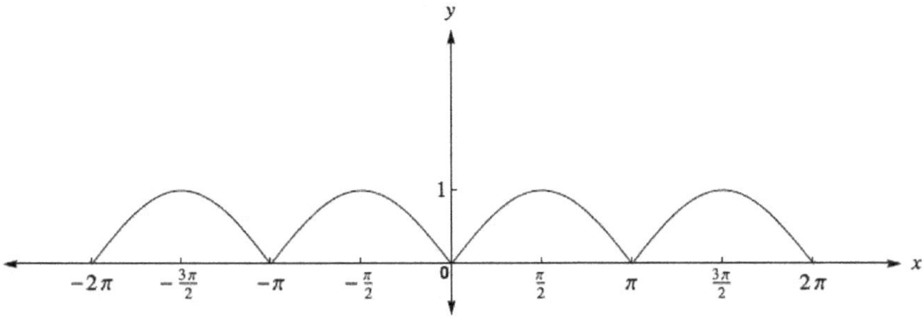

2. $|\csc x|$

Definition: Let x be a real number. Then the function h defined by

$$h(x) = |\csc x|, \csc x = 1/\sin x, \text{ and } \sin x \neq 0$$

is called the absolute value of the cosecant function.

The domain of h is {all real x: $x \neq \pm n\pi$, n = 0, 1, 2,...}, and the range of h is $[1, \infty)$. The vertical asymptotes of h are at $x = \pm n\pi$, n = 0, 1, 2 . . .

EXAMPLE: $h(\pi/6) = 2$, $h(\pi/2) = 1$, $h(5\pi/6) = 2$, $h(7\pi/6) = 2$, $h(-\pi/6) =2$, $h(-\pi/2) = 1$, $h(-5\pi/6) = 2$, and $h(-7\pi/6) = 2$. The graphs of $y = h(x)$ and $y = |\sin x|$ are shown in Figure 2.2:

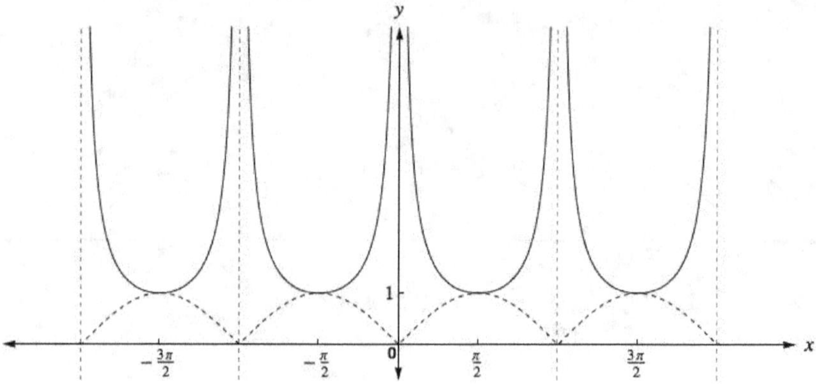

3. | cosx |

Definition: Let x be a real number. Then the function h defined by

$$h(x) = |\cos x|$$

is called the absolute value of the cosine function.

The domain of h is (-∞, ∞), and its range is [0, 1].

EXAMPLE: h(0) = 1, h(π/3) = 1/2, h(π/2) = 0, h(2π/3) = 1/2, h(π) = 1, h(-π/3) = 1/2, h(-π/2) = 0, h(-2π/3) = 1/2, h(-π) = 1.The graph of y = h(x) is shown in Figure 2.3:

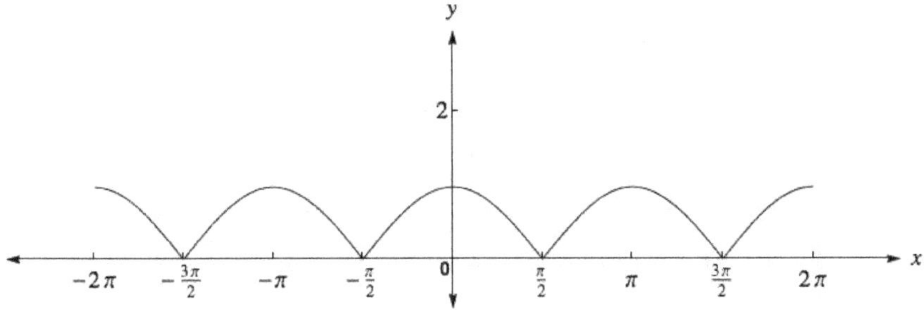

4. | secx |

Definition: Let x be a real number. Then the function h defined by

$$h(x) = |\sec x|, \ \sec x = 1/\cos x, \text{ and } \cos x \neq 0$$

is called the absolute value of the secant function.

The domain of h is {all real x: $x \neq \frac{\pi}{2} \pm n\pi$, n = 0, 1, 2 . . .}, and the range of h is $[1, \infty)$. The vertical asymptotes of h are at $x = \frac{\pi}{2} \pm n\pi$, n = 0, 1, 2 . . .

EXAMPLE: h(0) = 1, h(π/3) = 2, h(2π/3) = 2, h(π) = 1, h(-π/3) = 2, h(-2π/3) = 2, and h(-π) = 1. The graphs of y = h(x) and y = |cosx| are shown in Figure 2.4:

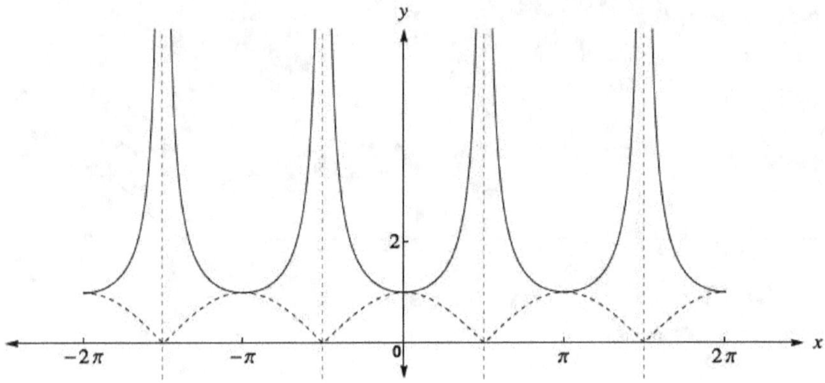

5. | tanx |

Definition: Let x be a real number. Then the function h defined by

$$h(x) = |\tan x|$$

is called the absolute value of the tangent function.

The domain of h is {all real x: $x \neq \frac{\pi}{2} \pm n\pi$, n = 0, 1, 2 . . .}, and the range of h is $[0, \infty)$. The vertical asymptotes of h are at $x = \frac{\pi}{2} \pm n\pi$, n = 0, 1, 2 . . .

EXAMPLE: h(0) = 0, h(π/4) = 1, h(3π/4) = 1, h(π) = 0, 1, h(-π/4) = 1, h(-3π/4) = 1, and h(-π) = 0. The graph of y = h(x) is shown in Figure 2.5:

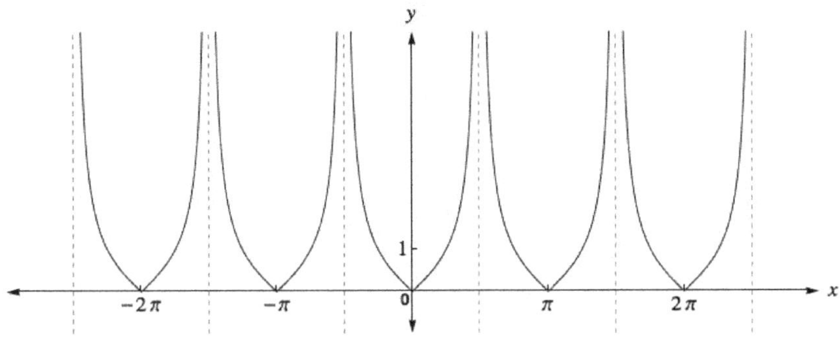

6. | cotx |

Definition: Let x be a real number. Then the function h defined by

$$h(x) = |\cot x|, \text{ and } \cot x = \tan(\frac{\pi}{2} - x)$$

is called the absolute value of the cotangent function.

The domain of h is {all real x: x ≠ ± nπ, n = 0, 1, 2 . . .}, and the range of h is [0, ∞). The vertical asymptotes of h are at x = ± nπ, n = 0, 1, 2 . . .

EXAMPLE: h(π/4) = 1, h(π/2) = 0, h(3π/4) = 1, h(-π/4) = 1, h(-π/2) = 0, and h(-3π/4) = 1. The graph of y = h(x) is shown in Figure 2.6:

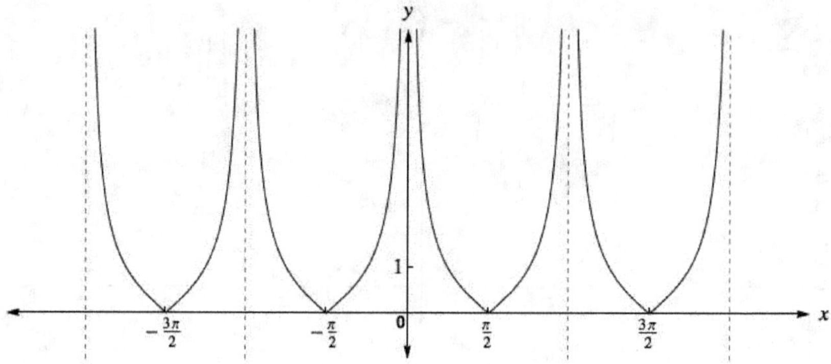
y
1
x
0
$-\frac{3\pi}{2}$
$-\frac{\pi}{2}$
$\frac{\pi}{2}$
$\frac{3\pi}{2}$

CHAPTER THREE

GRAPHS OF SINXTANX, COSXCOTX, SINXCOTX, COSXTANX, SINXCOSX, AND SECXCSCX

1. sinxtanx

Definition: Let x be a real number. Then the function h defined by

$$h(x) = \text{sinxtanx}$$

is called the product of the sine and tangent functions.

The domain of h is {all real x: $x \neq \frac{\pi}{2} \pm n\pi$, n = 0, 1, 2 . . .}, and the range of h is $(-\infty, \infty)$. The vertical asymptotes of h are at $x = \frac{\pi}{2} \pm n\pi$, n = 0, 1, 2 . . .

EXAMPLE: h(0) = 0, h(π/3) = 3/2, h(2π/3) = -3/2, h(π) = 0, h(-π/3) = 3/2, h(-2π/3) = -3/2, and h(-π) = 0. The graph of y = h(x) is shown in Figure 3.1:

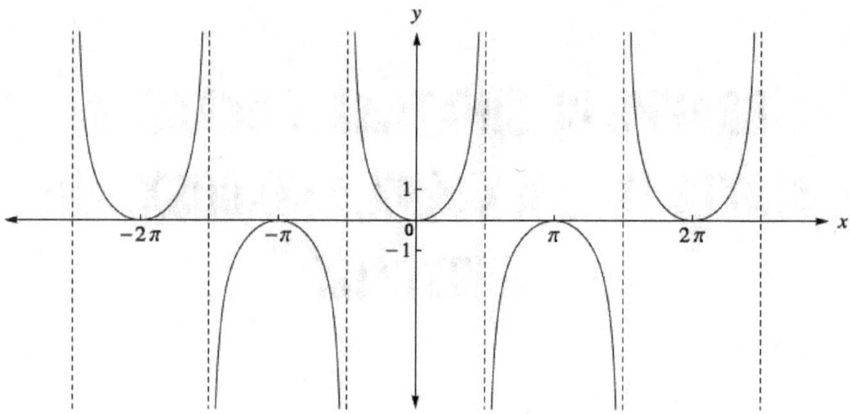

2. cosxcotx

Definition: Let x be a real number. Then the function h defined by

$$h(x) = \cos x \cot x, \text{ and } \cot x = \tan(\frac{\pi}{2} - x)$$

is called the product of the cosine and cotangent functions.

The domain of h is {all real x: x ≠ ± nπ, n = 0, 1, 2 . . .}, and the range of h is (-∞, ∞). The vertical asymptotes of h are at x = ± nπ, n = 0, 1, 2 . . .

EXAMPLE: h(π/6) = 3/2, h(π/2) = 0, h(5π/6) = 3/2, h(-π/6) = -3/2, h(-π/2) = 0, and h(-5π/6) = -3/2. The graph of y = h(x) is shown in Figure 3.2:

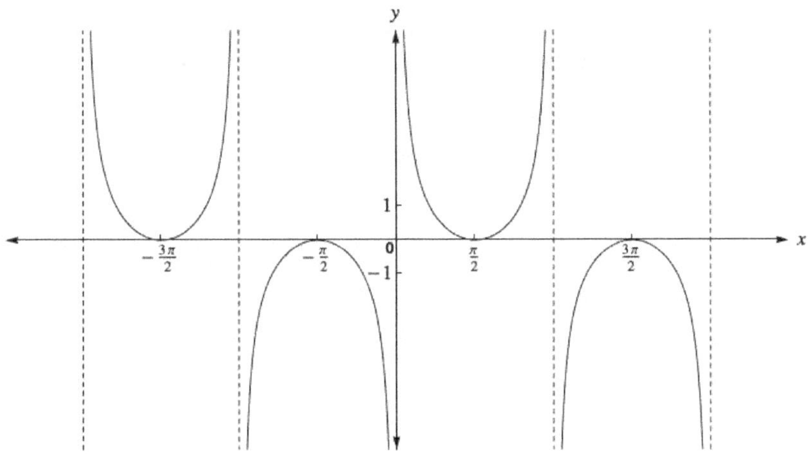

3. sinxcotx

Definition: Let x be a real number. Then the function h defined by

$$h(x) = \text{sinxcotx, and cotx} = \tan(\frac{\pi}{2} - x)$$

is called the product of the sine and cotangent functions.

The domain of h is {all real x: $x \neq \pm n\pi$, n = 0, 1, 2 . . .}, and the range of h is $(-\infty, \infty)$. The vertical asymptotes of h are at $x = \pm n\pi$, n = 0, 1, 2 . . .

EXAMPLE: $h(\pi/3) = 1/2$, $h(\pi/2) = 0$, $h(2\pi/3) = -1/2$, $h(-\pi/3) = 1/2$, $h(-\pi/2) = 0$, and $h(-2\pi/3) = -1/2$. The graph of $y = h(x)$ is shown in Figure 3.3:

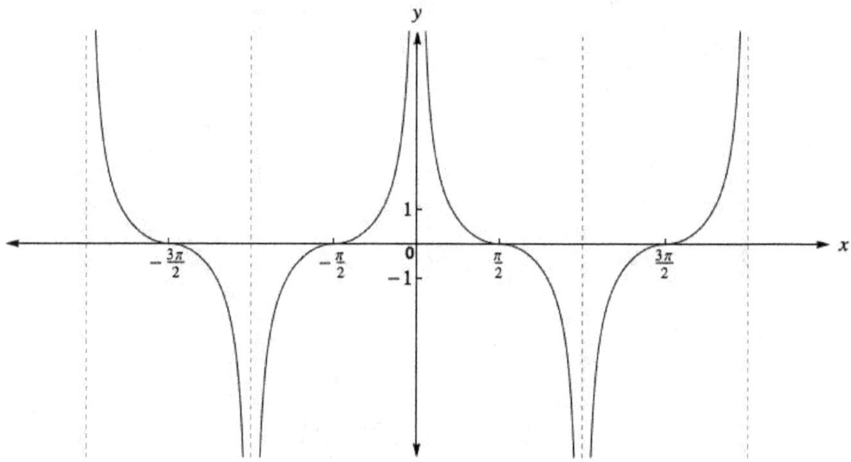

4. cosxtanx

Definition: Let x be a real number. Then the function h defined by

$$h(x) = \cos x \tan x$$

is called the product of the cosine and tangent functions.

The domain of h is {all real x: $x \neq \frac{\pi}{2} \pm n\pi$, n = 0, 1, 2 . . .}, and the range of h is $(-\infty, \infty)$. The vertical asymptotes of h are at $x = \frac{\pi}{2} \pm n\pi$, n = 0, 1, 2 . . .

EXAMPLE: h(0) = 0, h(π/6) = 1/2, h(5π/6) = 1/2, h(π) = 0, h(-π/6) = -1/2, h(-5π/6) = -1/2, and h(-π) = 0. The graph of y = h(x) is shown in Figure 3.4:

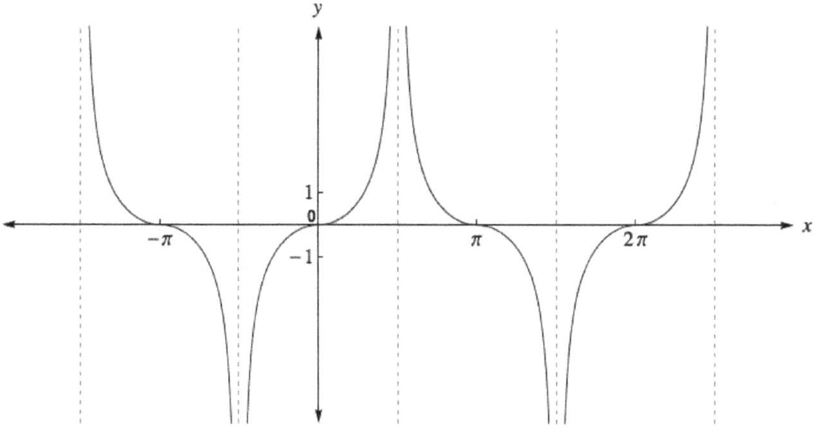

5. sinxcosx

Definition: Let x be a real number. Then the function h defined by

$$h(x) = \sin x \cos x$$

is called the product of the sine and cosine functions.

The domain of h is $(-\infty, \infty)$, and its range is $[-1/2, 1/2]$. The function h is continuous and periodic, and its period is π.

EXAMPLE: $h(0) = 0$, $h(\pi/4) = 1/2$, $h(3\pi/4) = -1/2$, $h(-\pi/4) = -1/2$, and $h(-3\pi/4) = 1/2$. The graph of $y = h(x)$ is shown in Figure 3.5:

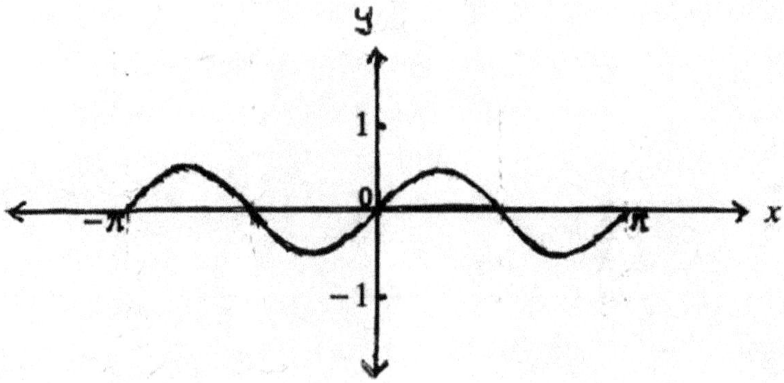

6. secxcscx

Definition: Let x be a real number. Then the function h defined by

$$h(x) = \sec x \csc x,\ \sec x \csc x = 1/(\sin x \cos x),\ \text{and } \sin x \cos x \neq 0$$

is called the product of the secant and cosecant functions.

The domain of h is {all real x: $x \neq \pm n\frac{\pi}{2}$, n = 0, 1, 2 . . .}, and the range of h is

(-∞, -2] ∪ [2, ∞). The vertical asymptotes of h are at $x = \pm n\frac{\pi}{2}$, n = 0, 1, 2 . . .

EXAMPLE: h(π/4) = 2, h(3π/4) = -2, h(-π/4) = -2, and h(-3π/4) = 2. The graphs of y = h(x) and y = sinxcosx are shown in Figure 3.6:

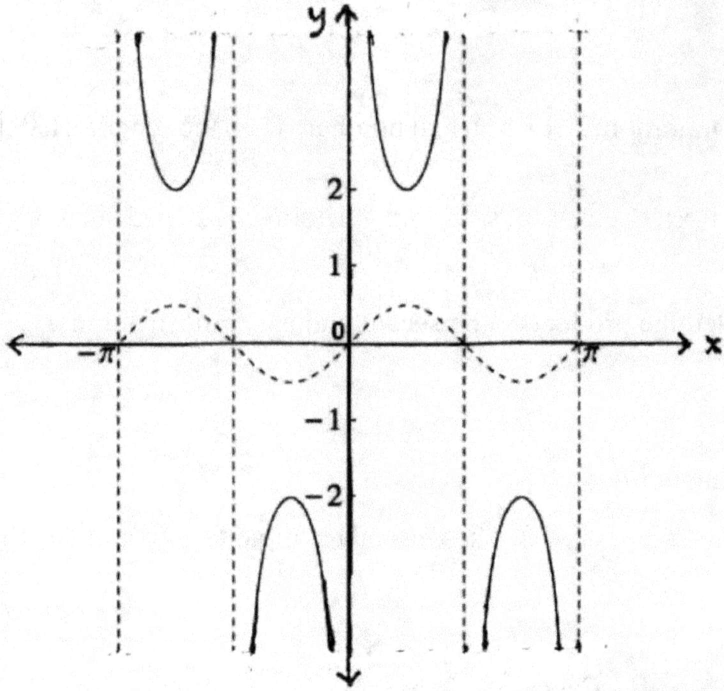
y
2
1
0
−1
−2
−π
π
x

CHAPTER FOUR

GRAPHS OF |SINX|TANX, SINX|TANX|, |COSX|TANX, COSX|TANX|, |SINX|COTX, SINX|COTX|, |COSX|COTX, AND COSX|COTX|

1. |sinx|tanx

Definition: Let x be a real number. Then the function h defined by

$$h(x) = |\sin x|\tan x$$

is called the product of the tangent function and the absolute value of the sine function.

The domain of h is {all real x: x $\neq (\frac{\pi}{2} \pm n\pi)$, n = 0, 1, 2 . . .}, and the range of h is $(-\infty, \infty)$. The vertical asymptotes of h are at x $= \frac{\pi}{2} \pm n\pi$, n = 0, 1, 2 . . .

EXAMPLE: h(0) = 0, h(π/3) = 3/2, h(2π/3) = -3/2, h(π) = 0, h(-π/3) = -3/2, h(-2π/3) = 3/2, and, h(-π) = 0. The graph of y = h(x) is shown in Figure 4.1:

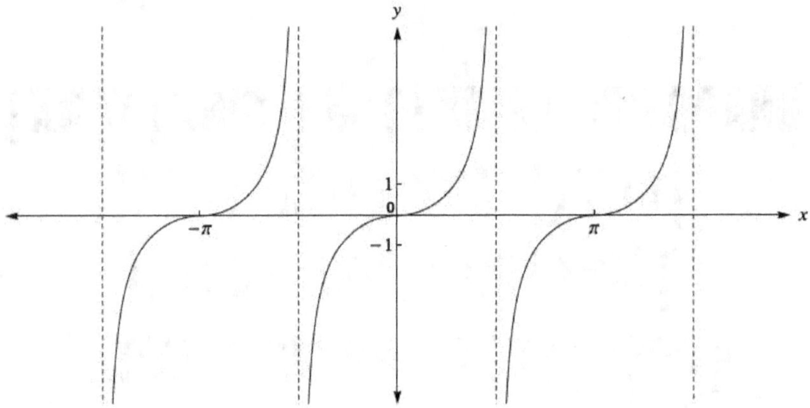

2. sinx | tanx |

Definition: Let x be a real number. Then the function h defined by

$$h(x) = \sin x|\tan x|$$

is called the product of the sine function and the absolute value of the tangent function

The domain of h is {all real x: $x \neq (\frac{\pi}{2} \pm n\pi)$, n = 0, 1, 2 . . .}, and the range of h is (-∞, ∞). The vertical asymptotes of h are at $x = \frac{\pi}{2} \pm n\pi$, n = 0, 1, 2 . . .

EXAMPLE: h(0) = 0, h(π/3) = 3/2, h(2π/3) = 3/2, h(π) = 0, h(-π/3) = -3/2, h(-2π/3) = -3/2, and h(-π) = 0. The graph of y = h(x) is shown in Figure 4.2:

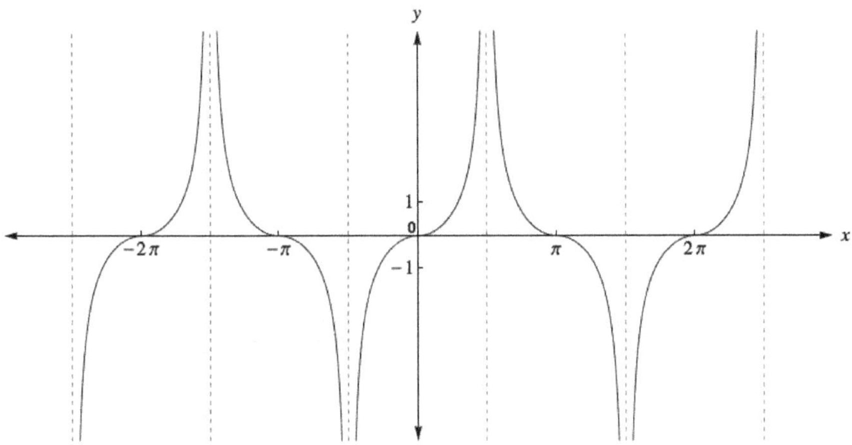

3. |cosx|tanx

Definition: Let x be a real number. Then the function h defined by

$$h(x) = |\cos x| \tan x$$

is called the product of the tangent function and the absolute value of the cosine function.

The domain of h is {all real x: $x \pm (\frac{\pi}{2} \pm n\pi)$, n = 0, 1, 2 . . .}, and the range of h is $(-\infty, \infty)$. The vertical asymptotes of h are at $x = \frac{\pi}{2} \pm n\pi$, n = 0, 1, 2 . . .

EXAMPLE: h(0) = 0, h(π/6) = 1/2, h(5π/6) = -1/2, h(π) = 0, h(-π/6) = -1/2, h(-5π/6) = 1/2, and h(-π) = 0. The graph of y = h(x) is shown in Figure 4.3:

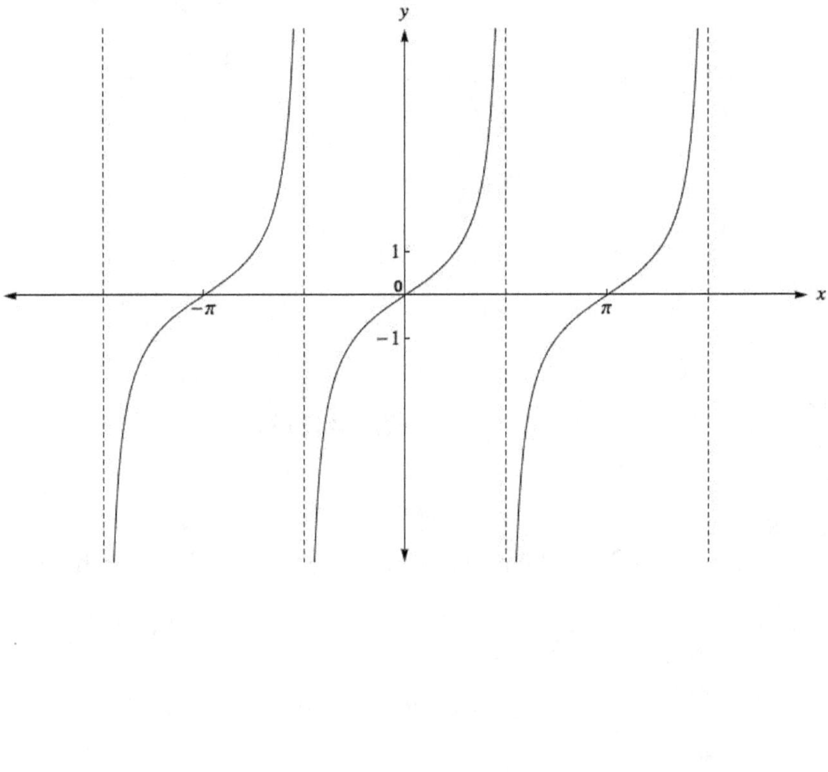

4. cosx | tanx |

Definition: Let x be a real number. Then the function h defined by

$$h(x) = \cos x|\tan x|$$

is called the product of the cosine function and the absolute value of the tangent function.

The domain of h is {all real x: x ≠ ($\frac{\pi}{2}$ ± nπ), n = 0, 1, 2 . . .}, and the range of h is (-∞, ∞). The vertical asymptotes of h are at x = $\frac{\pi}{2}$ ± nπ, n = 0, 1, 2 . . .

EXAMPLE: h(0) = 0, h(π/6) = 1/2, h(5π/6) = -1/2, h(π) = 0, h(-π/6) = 1/2, h(-5π/6) = -1/2, and h(-π) = 0. The graph of y = h(x) is shown in Figure 4.4:

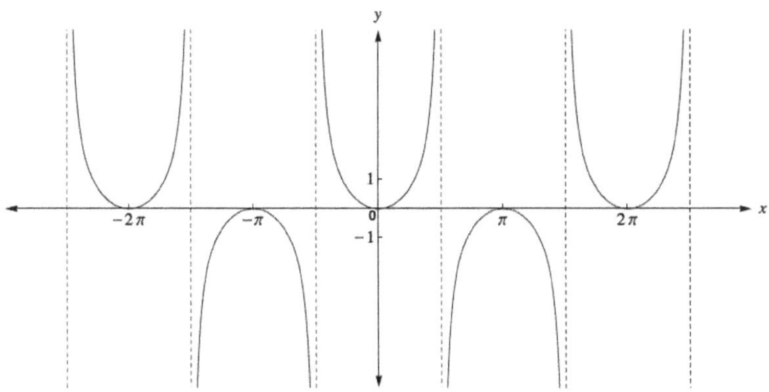

5. $|\sin x|\cot x$

Definition: Let x be a real number. Then the function h defined by

$$h(x) = |\sin x| \cot x, \text{ and } \cot x = \tan(\tfrac{\pi}{2} - x)$$

is called the product of the cotangent function and the absolute value of the sine function.

The domain of h is {all real x: $x \neq \pm n\pi$, n = 0, 1, 2 . . .}, and the range of h is $(-\infty, \infty)$. The vertical asymptotes of h are at $x = \pm n\pi$, n = 0, 1, 2 . . .

EXAMPLE: h(π/3) = 1/2, h(π/2) = 0, h(2π/3) = -1/2, h(-π/3) = -1/2, h(-π/2) = 0, and h(-2π/3) = 1/2. The graph of y = h(x) is shown in Figure 4.5:

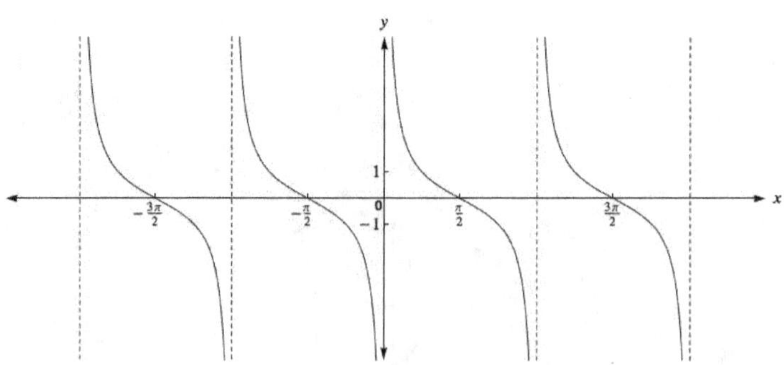

6. sinx | cotx |

Definition: Let x be a real number. Then the function h defined by

$$h(x) = \sin x \,|\cot x|, \text{ and } \cot x = \tan\left(\frac{\pi}{2} - x\right)$$

is called the product of the sine function and the absolute value of the cotangent function.

The domain of h is {all real x: x ≠ ± nπ, n = 0, 1, 2 . . .}, and the range of h is (-∞, ∞). The vertical asymptotes of h are at x = ± nπ, n = 0, 1, 2 . . .

EXAMPLE: h(π/3) = 1/2, h(π/2) = 0, h(2π/3) = 1/2, h(-π/3) = -1/2, h(-π/2) = 0, and h(-2π/3) = -1/2. The graph of y = h(x) is shown in Figure 4.6:

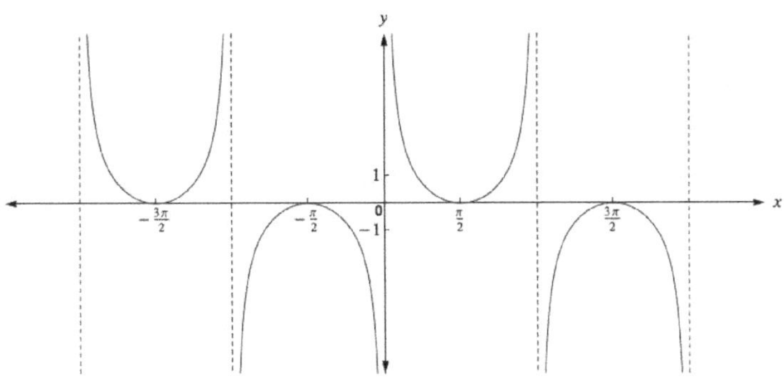

7. $|\cos x| \cot x$

Definition: Let x be a real number. Then the function h defined by

$$h(x) = |\cos x| \cot x, \text{ and } \cot x = \tan\left(\frac{\pi}{2} - x\right)$$

is called the product of the cotangent function and the absolute value of the cosine function.

The domain of h is {all real x: $x \neq \pm n\pi$, n = 0, 1, 2 . . .}, and the range of h is $(-\infty, \infty)$. The vertical asymptotes of are at $x = \pm n\pi$, n = 0, 1, 2 . . .

EXAMPLE: $h(\pi/6) = 3/2$, $h(\pi/2) = 0$, $h(5\pi/6) = -3/2$, $h(-\pi/6) = -3/2$, $h(-\pi/2) = 0$, and $h(-5\pi/6) = 3/2$. The graph of $y = h(x)$ is shown in Figure 4.7:

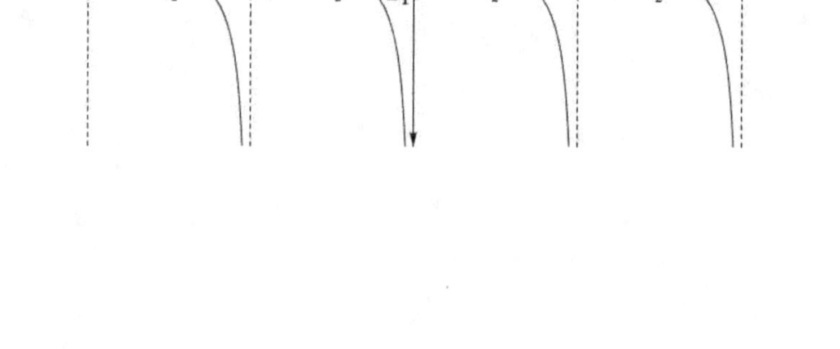

8. cosx | cotx |

Definition: Let x be a real number. Then the function h defined by

$$h(x) = \cos x\,|\cot x|, \text{ and } \cot x = \tan(\tfrac{\pi}{2} - x)$$

is called the product of the cosine function and the absolute value of the cotangent function.

The domain of h is {all real x: x ≠ ± nπ, n = 0, 1, 2 . . .}, and the range of h is (-∞, ∞). The vertical asymptotes of h are at x = ± nπ, n = 0, 1, 2 . . .

EXAMPLE: h(π/6) = 3/2, h(π/2) = 0, h(5π/6) = -3/2, h(-π/6) = 3/2, h(-π/2) = 0, and h(-5π/6) = -3/2. The graph of y = h(x) is shown in Figure 4.8:

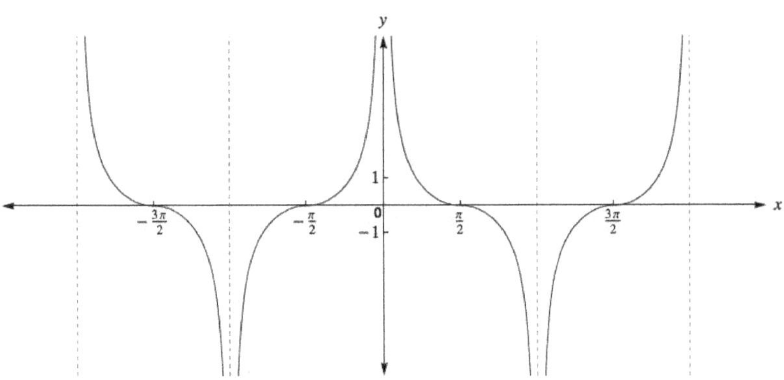

CHAPTER FIVE

GRAPHS OF SINXTAN | X | , COSXTAN | X | , TANXTAN | X | , SINXCOT | X | , COSXCOT | X |, AND COTXCOT | X |

1. sinxtan | x |

Definition A: Let x be a real number. Then the function h defined by

$$h(x) = \sin x \tan|x|$$

is called the product of the sine function and the tangent of the absolute value function.

The domain of h is {all real x: $x \neq \frac{\pi}{2} \pm n\pi$, n = 0, 1, 2 . . .}, and the range of h is $(-\infty, \infty)$. The vertical asymptotes of h are at $x = \frac{\pi}{2} \pm n\pi$, n = 0, 1, 2 . . .

EXAMPLE A: $h(0) = 0$, $h(\pi/4) = 1/\sqrt{2}$, $h(3\pi/4) = -1/\sqrt{2}$, $h(\pi) = 0$, $h(-\pi/4) = -1/\sqrt{2}$, $h(-3\pi/4) = 1/\sqrt{2}$, and $h(-\pi) = 0$. The graph of $y = h(x)$ is shown in Figure 5.1:

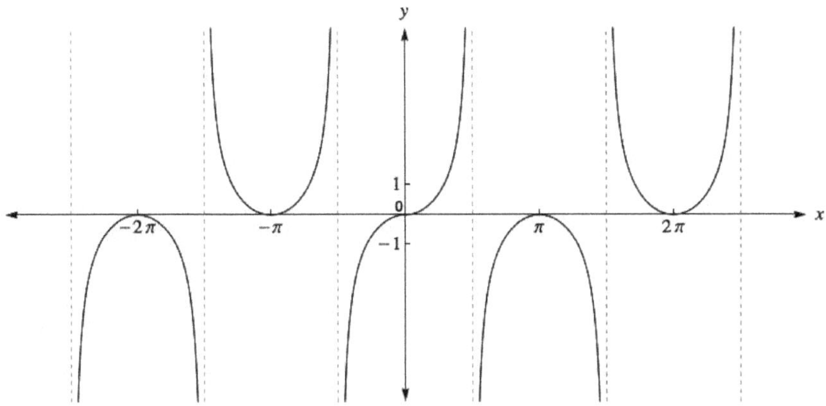

2. cosxtan|x|

Definition: Let x be a real number. Then the function h defined by

$$h(x) = \cos x \tan|x|$$

is called the product of the cosine function and the tangent of the absolute value function.

The domain of h is {all real x: $x \neq \frac{\pi}{2} \pm n\pi$, n = 0, 1, 2 . . .}, and the range of h is $(-\infty, \infty)$. The vertical asymptotes of h are at $x = \frac{\pi}{2} \pm n\pi$, n = 0, 1, 2 . . .

EXAMPLE: h(0) = 0, h(π/4) = 1/ $\sqrt{2}$, h(3π/4) = 1/ $\sqrt{2}$, h(π) = 0, h(5π/4) = -1/ $\sqrt{2}$, h(7π/4) = -1/ $\sqrt{2}$, h(-π/4) = 1/ $\sqrt{2}$, h(-3π/4) = 1/ $\sqrt{2}$, h(-π) = 0, h(-5π/4) = -1/ $\sqrt{2}$, and h(-7π/4) = -1/ $\sqrt{2}$. The graph of y = h(x) is show in Figure 5.2:

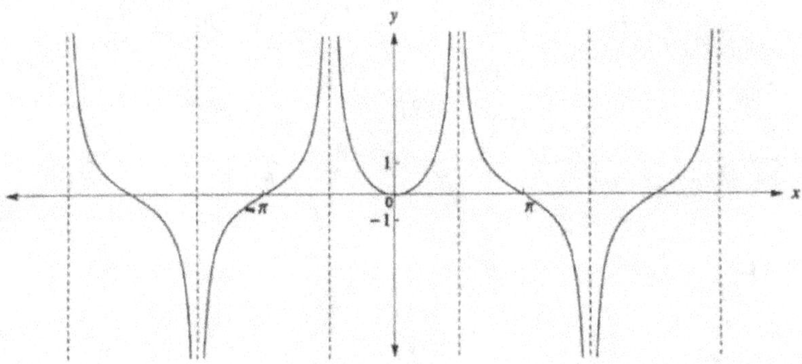

3. tanxtan | x |

Definition: Let x be a real number. Then the function h defined by

$$h(x) = \tan x \tan |x|$$

is called the product of the tangent function and the tangent of the absolute value function.

The domain of h is {all real x: $x \neq \frac{\pi}{2} \pm n\pi$, n = 0, 1, 2 . . .}, and the range of h is (-∞, ∞). The vertical asymptotes of h are at $x = \frac{\pi}{2} \pm n\pi$, n = 0, 1, 2 . . .

EXAMPLE: h(0) = 0, h(π/4) = 1, h(3π/4) = 1, h(π) = 0, h(5π/4) = 1, h(-π/4) = -1, h(-3π/4) = -1, h(-π) = 0, and h(-5π/4) = -1. The graph of y = h(x) is shown in Figure 5.3:

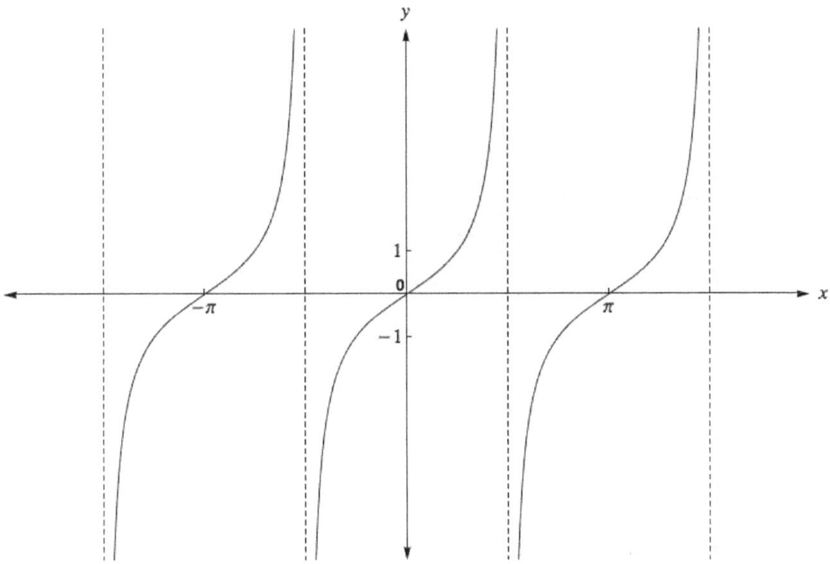

4. sinxcot|x|

Definition: Let x be a real number. Then the function h defined by

$$h(x) = \sin x \cot|x|, \text{ and } \cot|x| = \tan(\frac{\pi}{2} - |x|)$$

is called the product of the sine function and the cotangent of the absolute value function.

The domain of h is {all real x: $x \neq \pm n\pi$, n = 0, 1, 2 . . .}, and the range of h is $(-\infty, \infty)$. The vertical asymptotes of h are at $x = \pm n\pi$, n = 0, 1, 2 . . .

EXAMLE: $h(\pi/4) = 1/\sqrt{2}$, $h(\pi/2) = 0$, $h(3\pi/) = -1/\sqrt{2}$, $h(5\pi/4) = -1/\sqrt{2}$, $h(3\pi/2) = 0$, $h(7\pi/4) = 1/\sqrt{2}$, $h(9\pi/4) = 1/\sqrt{2}$, $h(5\pi/2) = 0$, $h(-\pi/4) = -1/\sqrt{2}$, $h(-\pi/2) = 0$, $h(-3\pi/4) = 1/\sqrt{2}$, $h(-5\pi/4) = 1/\sqrt{2}$, $h(-3\pi/2) = 0$, $h(-7\pi/4) = -1/\sqrt{2}$, $h(-9\pi/4) = -1/\sqrt{2}$, and $h(-5\pi/2) = 0$. The graph of y = h(x) is shown in Figure 5.4:

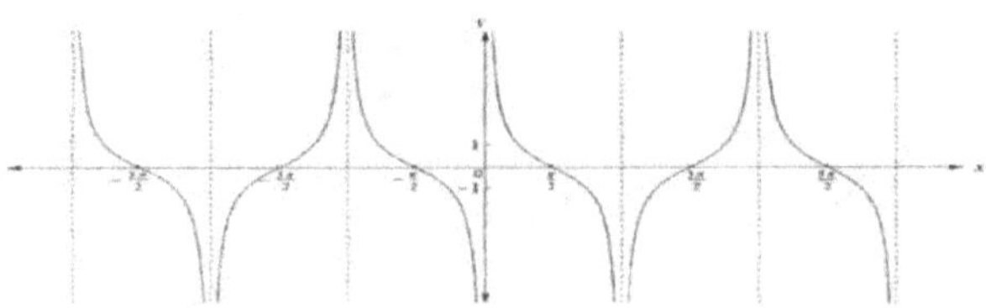

5. cosxcot|x|

Definition: Let x be a real number. Then the function h defined by

$$h(x) = \cos x \cot|x|, \text{ and } \cot|x| = \tan(\tfrac{\pi}{2} - |x|)$$

is called the product of the cosine function and the cotangent of the absolute value function.

The domain of h is {all real x: $x \neq \pm n\pi$, n = 0, 1, 2 . . . }, and the range of h is $(-\infty, \infty)$.

EXAMPLE: $h(\pi/4) = 1/\sqrt{2}$, $h(3\pi/4) = 1/\sqrt{2}$, $h(5\pi/4) = -1/\sqrt{2}$, $h(-\pi/4) = 1/\sqrt{2}$, $h(-3\pi/4) = 1/\sqrt{2}$, and $h(-5\pi/4) = -1/\sqrt{2}$. The graph of y = h(x) is shown in Figure 5.5:

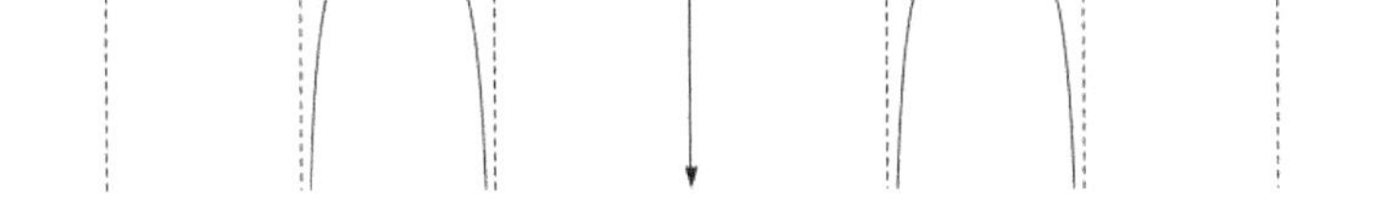

6. cotxcot|x|

Definition: Let x be a real number. Then the function h defined by

$h(x) = \cot x \cot|x|$, $\cot x = \tan(\frac{\pi}{2} - x)$, and $\cot|x| = \tan(\frac{\pi}{2} - |x|)$

is called the product of the cotangent function and the cotangent of the absolute value function.

The domain of h is {all real x: $x \neq \pm n\pi$, n = 0, 1, 2 . . .}, and the range of h is $(-\infty, \infty)$. The vertical asymptotes of h are at $x = \pm n\pi$, n = 0, 1, 2 . . .

EXAMPLE: h(π/4) = 1, h(π/2) = 0, h(3π/4) = 1, h(3π/2) = 0, h(-π/4) = -1, h(-π/2) = 0, h(-3π/4) = -1, and h(-3π/2) = 0. The graph of y = h(x) is shown in Figure 5.6:

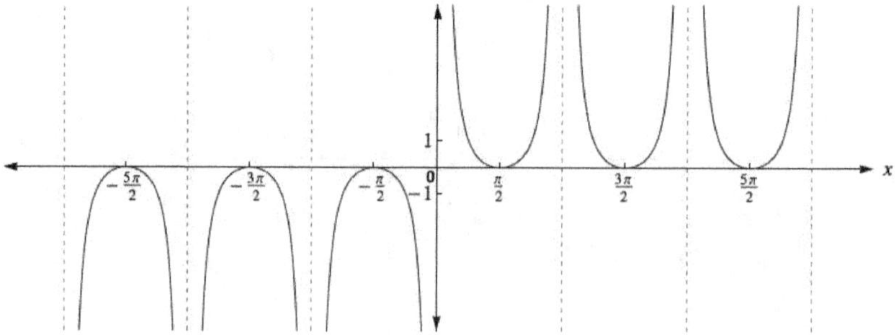

CHAPTER SIX

GRAPHS OF SIN | X | COSX, SIN | X | | COSX | , COSX | SINX | , SINX | COSX | , AND SIN | X | SINX

1. sin | x | cosx

Definition: Let x be a real number. Then the function h defined by

$$h(x) = \sin|x|\cos x$$

is called the product of the cosine function and the sine of the absolute value function.

The domain of h is $(-\infty, \infty)$, and its range is $[-1/2, 1/2]$.

EXAMPLE: $h(0) = 0$, $h(\pi/4) = 1/2$, $h(\pi/2) = 0$, $h(3\pi/4) = -1/2$, $h(\pi) = 0$, $h(-\pi/4) = 1/2$, $h(-\pi/2) = 0$, $h(-3\pi/4) = -1/2$, and $h(-\pi) = 0$. The graph of $y = h(x)$ is shown in Figure 6.1:

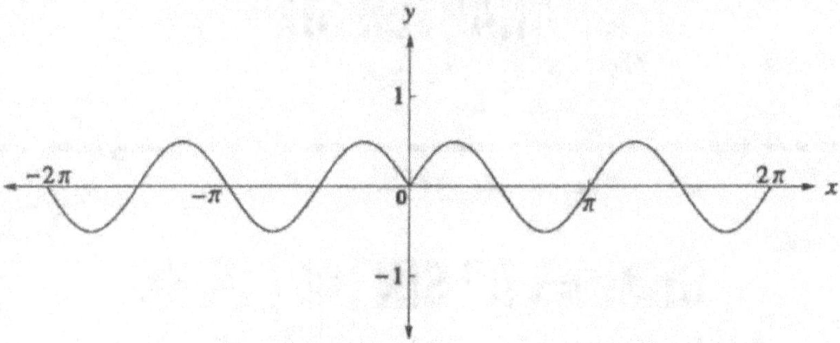

2. sin|x| |cosx|

Definition: Let x be a real number. Then the function h defined by

$$h(x) = \sin|x||\cos x|$$

is called the product of the absolute value of the cosine function and the sine of the absolute value function.

The domain of h is $(-\infty, \infty)$, and its range is $[-1/2, 1/2]$.

EXAMPLE: $h(0) = 0$, $h(\pi/4) = 1/2$, $h(\pi/2) = 0$, $h(3\pi/4) = 1/2$, $h(\pi) = 0$, $h(5\pi/4) = -1/2$, $h(3\pi/2) = 0$, $h(7\pi/4) = -1/2$, $h(2\pi) = 0$, $h(-\pi/4) = 1/2$, $h(-\pi/2) = 0$, $h(-3\pi/4) = 1/2$, $h(-\pi) = 0$, $h(-5\pi/4) = -1/2$, $h(-3\pi/2) = 0$, $h(-7\pi/4) = -1/2$, and $h(-2\pi) = 0$. The graph of $y = h(x)$ is shown in Figure 6.2:

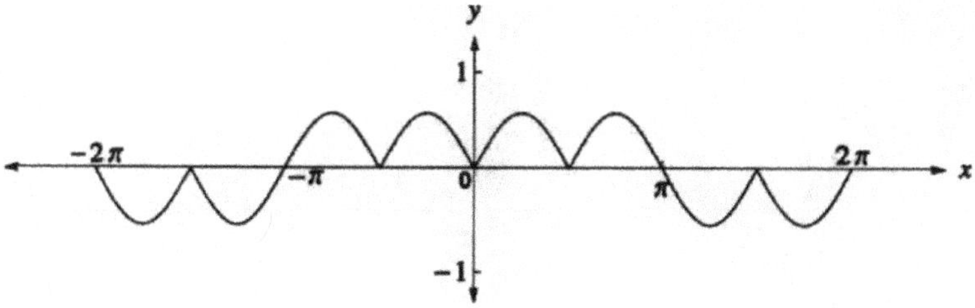

3. cosx | sinx |

Definition: Let x be a real number. Then the function h defined by

$$h(x) = \cos x|\sin x|$$

is called the product of the cosine function and the absolute value of the sine function.

The domain of h is (-∞, ∞), and its range is [-1/2, 1/2].

EXAMPLE: h(0) = 0, h(π/4) = 1/2, h(π/2) = 0, h(3π/4) = -1/2, h(π) = 0, h(5π/4) = -1/2, h(3π/2) = 0, h(7π/4) = 1/2, h(2π) = 0, h(-π/4) = 1/2, h(-π/2) = 0, h(-3π/4) = -1/2, h(-π) = 0, h(-5π/4) = -1/2, h(-3π/2) = 0, h(-7π/4) = 1/2, and h(-2π) = 0. The graph of y = h(x) is shown in Figure 6.3:

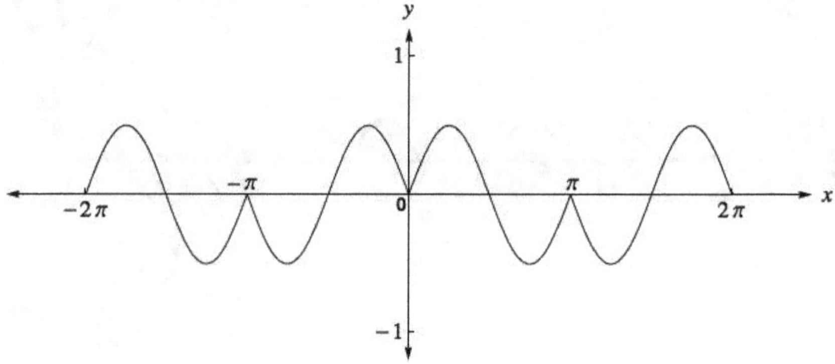

4. sinx | cosx |

Definition: Let x be a real number. Then the function h defined by

$$h(x) = \sin x|\cos x|$$

is called the product of the sine function and the absolute value of the cosine function.

The domain of h is $(-\infty, \infty)$, and its range is $[-1/2, 1/2]$.

EXAMPLE: $h(0) = 0$, $h(\pi/4) = 1/2$, $h(\pi/2) = 0$, $h(3\pi/4) = 1/2$, $h(\pi) = 0$, $h(5\pi/4) = -1/2$, $h(3\pi/2) = 0$, $h(7\pi/4) = -1/2$, $h(2\pi) = 0$, $h(-\pi/4) = -1/2$, $h(-\pi/2) = 0$, $h(-3\pi/4) = -1/2$, $h(-\pi) = 0$, $h(-5\pi/4) = 1/2$, $h(-3\pi/2) = 0$, $h(-7\pi/4) = 1/2$, and $h(-2\pi) = 0$. The graph of $y = h(x)$ is shown in Figure 6.4:

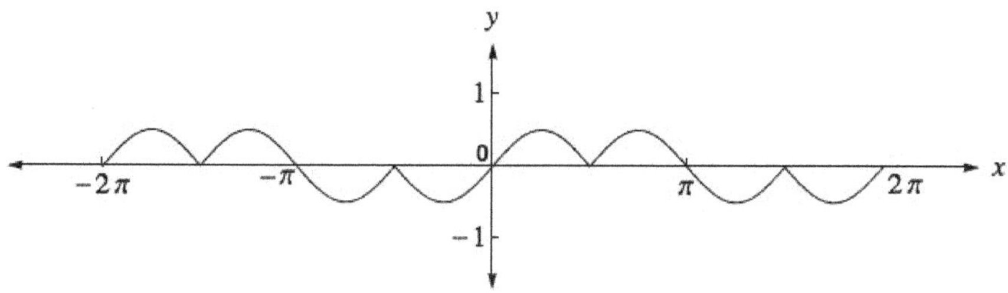

5. sin | x | sinx

Definition: Let x be a real number. Then the function h defined by

$$h(x) = \sin|x|\sin x$$

is called the product of the sine function and the sine of the absolute value function.

The domain of h is (-∞, ∞), and its range is [-1, 1].

EXAMPLE: h(0) = 0, h(π/4) = 1/2, h(π/2) = 1, h(3π/4) = 1/2, h(π) = 0, h(5π/4) = 1/2, h(3π/2) = 1, h(7π/4) = 1/2, h(2π) = 0, h(-π/4) = -1/2, h(-π/2) = -1, h(-3π/4) = -1/2, h(-π) = 0, h(-5π/4) = -1/2, h(-3π/2) = -1, h(-7π/4) = -1/2, and h(-2π) = 0. The graph of y = h(x) is shown in Figure 6.5:

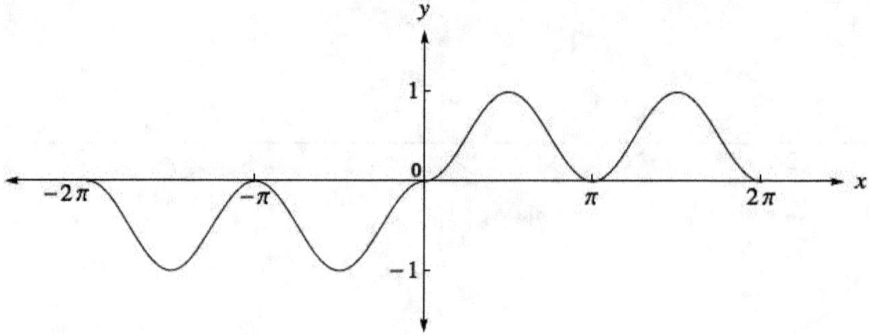
y
1
0
−1
x
−2π
−π
π
2π

CHAPTER SEVEN

GRAPHS OF CSC | X | SECX, CSC | X | | SECX |, SECX | CSCX | ,CSCX | SECX |, AND CSC | X | CSCX

1. csc | x | secx

Definition: Let x be a real number. Then the function h defined by

$$h(x) = \csc|x|\sec x,\ \csc|x|\sec x = 1/(\sin|x|\cos x), \text{ and } \sin|x|\cos x \neq 0$$

is called the product of the secant function and the cosecant of the absolute value function.

The domain of h is {all real x: $x \neq \pm\frac{\pi}{2}n$, n = 0, 1, 2 . . .}, and the range of h is $(-\infty, -2] \cup [2, \infty)$. The vertical asymptotes of h are at $x = \pm\frac{\pi}{2}n$, n = 0, 1, 2 . . .

The domain of sin|x|cosx is (-∞, ∞), and its range is [-1/2, 1/2].

EXAMPLE: h(π/4) = 2, h(3π/4) = -2, h(5π/4) = 2, h(-π/4), h(-3π/4) = -2, and h(-5π/4) = 2. The graphs of y = h(x) and y = sin|x|cosx are shown in Figure 7.1:

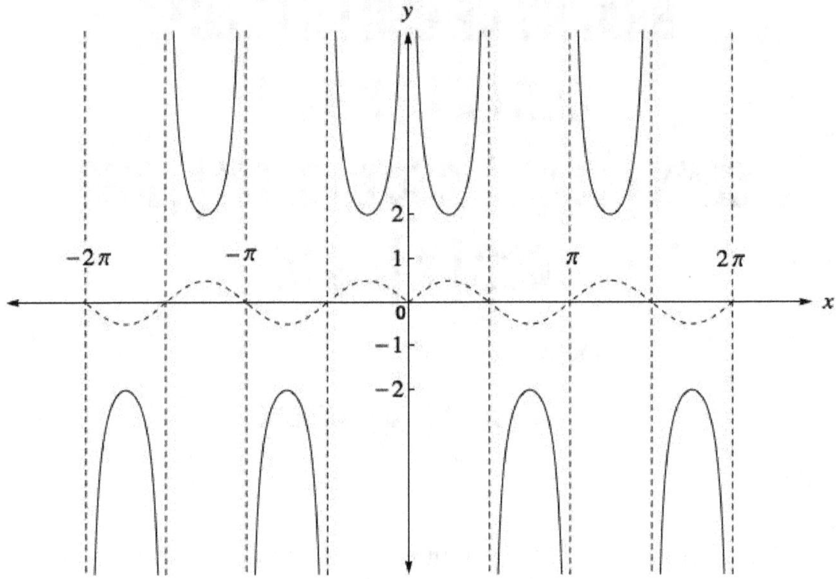

2. csc|x| |secx|

Definition: Let x be a real number. Then the function h defined by

h(x) = csc|x||secx|, csc|x||secx|, 1/(sin|x||cosx|), and sin||x||cosx| ≠ 0

is called the product of the absolute value of the secant function and the cosecant of the absolute value function.

The domain of h is {all real x: x $\neq \pm\frac{\pi}{2}$ n, n = 0, 1, 2 . . .}, and the range of h is (-∞, -2] ∪ [2, ∞). The vertical asymptotes of h are at x = $\pm\frac{\pi}{2}$ n, n = 0, 1, 2 . . .

The domain of sin|x||cosx| is (-∞, ∞), and its range is [-1/2, 1/2].

EXAMPLE: h(π/4) = 2, h(3π/4) = 2, h(5π/4) = -2, h(7π/4) = -2, h(-π/4) = 2, h(-3π/4) = 2, h(-5π/4) = -2, and h(-7π/4) = -2. The graphs of y = h(x) and y = sin|x||cosx| are shown in Figure 7.2:

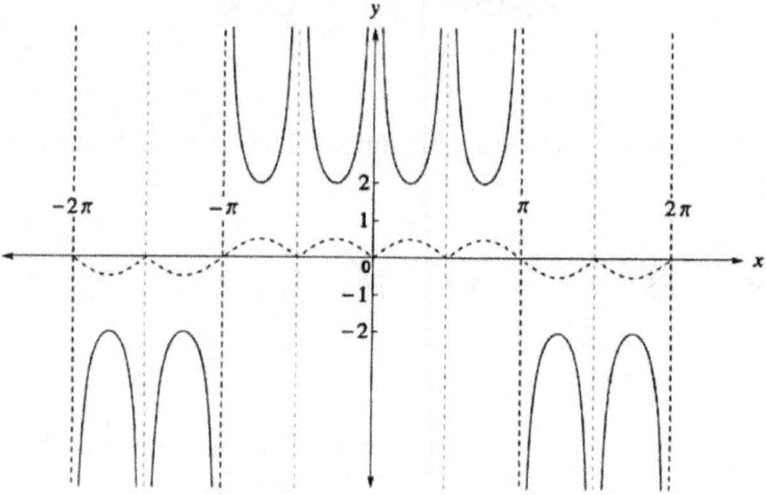

3. secx | cscx |

Definition: Let x be a real number. Then the function h defined by

$$h(x) = \sec x|\csc x|,\ \sec x|\csc x| = 1/(\cos x|\sin x|),\ \text{and } \cos x|\sin x| \neq 0$$

is called the product of the secant function and the absolute value of the cosecant function.

The domain of h is {all real x: $x \neq \pm\frac{\pi}{2}n$, n = 0, 1, 2,...}, and the range of h is $(-\infty, -2] \cup [2, \infty)$. The vertical asymptotes of h are at $x = \pm\frac{\pi}{2}n$, n = 0, 1, 2,...

The domain of cosx|sinx| is (-∞, ∞), and its range is [-1/2, 1/2].

EXAMPLE: h(π/4) = 2, h(3π/4) = -2, h(5π/4) = -2, h(7π/4) = 2, h(-π/4) = 2, h(-3π/4) = -2, h(-5π/4) = -2, and h(-7π/4) = 2. The graphs of y = h(x) and y = cosx|sinx| are shown in Figure 7.3.

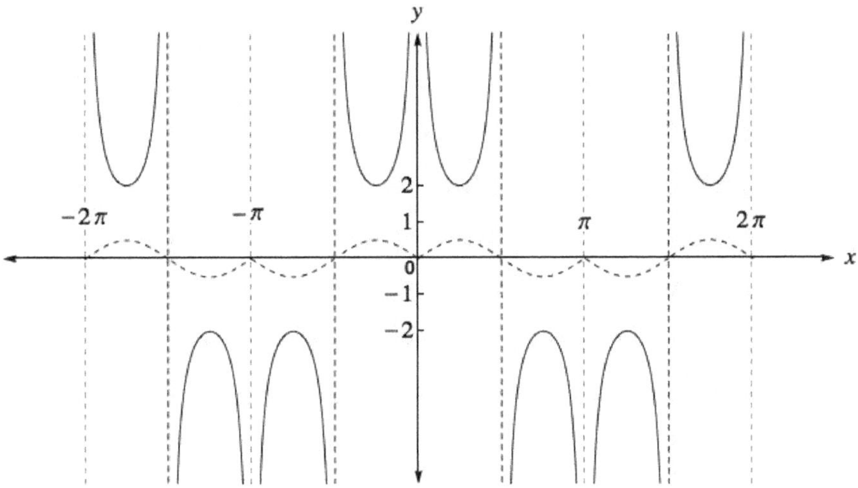

4. cscx|secx|

Definition: Let x be a real number. Then the function h defined by

$$h(x) = \csc x|\sec x|,\ \csc x|\sec x| = 1/(\sin x|\cos x|),\text{ and } \sin x|\cos x| \neq 0$$

is called the product of the cosecant function and the absolute value of the secant function.

The domain of h is {all real x: $x \neq \pm\frac{\pi}{2}n$, n = 0, 1, 2,...}, and the range of h is $(-\infty, -2] \cup [2, \infty)$. The vertical asymptotes of h are at $x = \pm\frac{\pi}{2}n$, n = 0, 1, 2,...

The domain of sinx|cosx| is $(-\infty, \infty)$, and its range is [-1/2, 1/2].

EXAMPLE: h(π/4) = 2, h(3π/4) = 2, h(5π/4) = -2, h(7π/4) = -2, h(-π/4) = -2, h(-3π/4) = -2, h(-5π/4) = 2 and h(-7π/4) = 2. The graphs of y = h(x) and y = sinx|cosx| are shown in Figure 7.4:

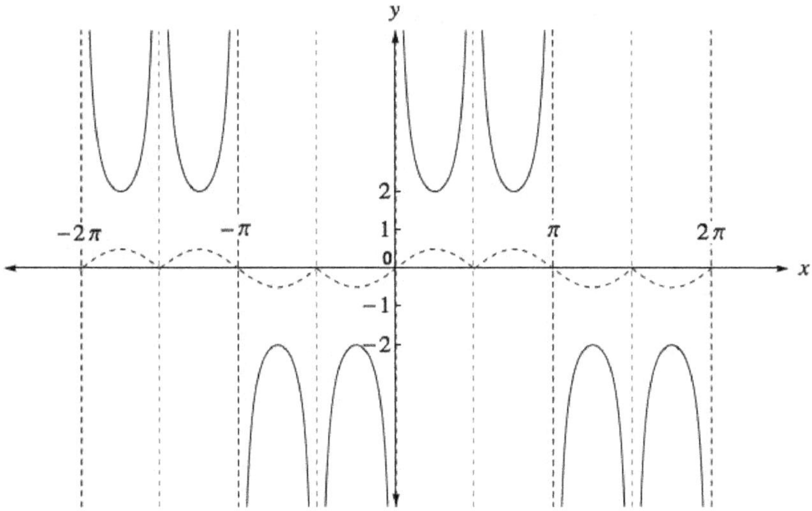

5. csc | x | cscx

Definition: Let x be a real number. Then the function h defined by

$$h(x) = \csc|x|\csc x, \ \csc|x|\csc x = 1/(\sin|x|\sin x), \text{ and } \sin|x|\sin x \neq 0$$

is called the product of the cosecant function and the cosecant of the absolute value function.

The domain of h is {all real x: $x \neq \pm n\pi$, n = 0, 1, 2 . . .}, and the range of h is $(-\infty, -1) \cup [1, \infty]$. The vertical asymptotes of h are at $x = \pm n\pi$, n = 0, 1, 2 . . .

EXAMPLE: $h(\pi/2) = 1$, $h(3\pi/2) = 1$, $h(-\pi/2) = -1$, and $h(-3\pi/2) = -1$. The graphs of $y = h(x)$ and $y = \sin|x|\sin x$ are shown in Figure 7.5:

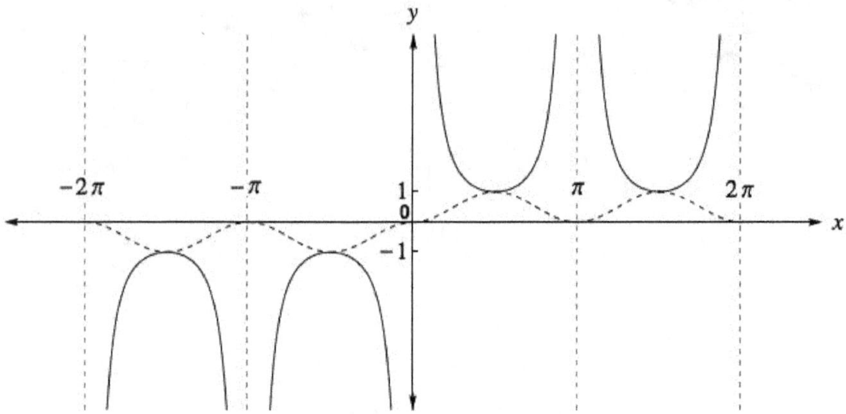

CHAPTER EIGHT

GRAPHS OF |SINXCOSX|, |SECXCSCX|, |SIN|X|SINX|, AND |CSC|X|CSCX|

1. |sinxcosx|

Definition: Let x be a real number. Then the function h defined by

$$h(x) = |\sin x \cos x|$$

is called the absolute value of the product of the sine and cosine functions.

The domain of h is $(-\infty, \infty)$, and its range is $[0, \frac{1}{2}]$. The period of h is $\pi/2$.

The domain of sinxcosx is $(-\infty, \infty)$, and its range is $[-1/2, \frac{1}{2}]$. The period of sinxcosx is π.

EXAMPLE: $h(0) = 0$, $h(\pi/6) = \sqrt{3}/4$, $h(\pi/4) = \frac{1}{2}$, $h(\pi/3) = \sqrt{3}/4$, $h(\pi/2) = 0$, $h(2\pi/3) = \sqrt{3}/4$, $h(3\pi/4) = \frac{1}{2}$, $h(5\pi/6) = \sqrt{3}/4$, $h(\pi)$

$= 0$, h(-π/6) = $\sqrt{3}$ /4, h(-π/4) = ½, h(-π/3) = $\sqrt{3}$ /4, h(-π/2) = 0, h(-2π/3) = $\sqrt{3}$ /4, h(-3π/4) = ½, h(-5π/6) = $\sqrt{3}$ /4, and h(-π) = 0. A part of the graph of y = h(x) is shown in Figure 8.1:

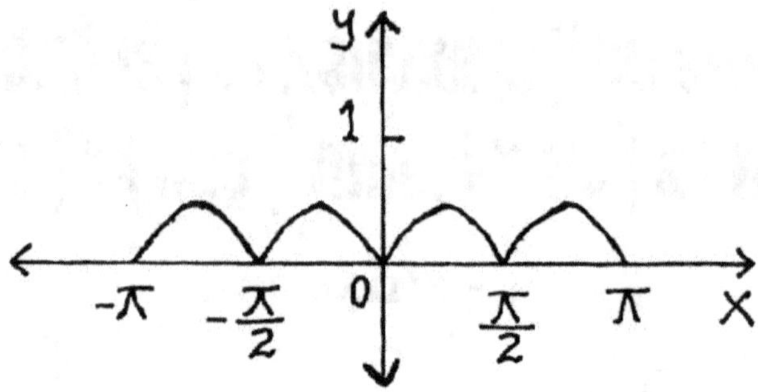

2. |secxcscx|

Definition: Let x be a real number. Then the function h defined by

h(x) = |secxcscx|, secxcscx = 1/(sinxcosx), and sinxcosx ≠ 0

is called the absolute value of the product of the secant and cosecant functions.

The domain of h is {all real x: x ≠ ± $\frac{\pi}{2}$ n, n = 0, 1, 2 . . .}, and the range of h is [2, ∞]. The vertical asymptotes of h are at x = ± $\frac{\pi}{2}$ n, n = 0, 1, 2 . . .

EXAMPLE: $h(\pi/6) = 4/\sqrt{3}$, $h(\pi/4) = 2$, $h(\pi/3) = 4/\sqrt{3}$, $h(2\pi/3) = 4/\sqrt{3}$, $h(3\pi/4) = 2$, $h(5\pi/6) = 4/\sqrt{3}$, $h(-\pi/6) = 4/\sqrt{3}$, $h(-\pi/4) = 2$, $h(-\pi/3) = 4/\sqrt{3}$, $h(-2\pi/3) = 4/\sqrt{3}$, $h(-3\pi/4) = 2$, and $h(-5\pi/6) = 4/\sqrt{3}$. The graphs of $y = h(x)$ and $y = |\sin x \cos x|$ are shown in Figure 8.2:

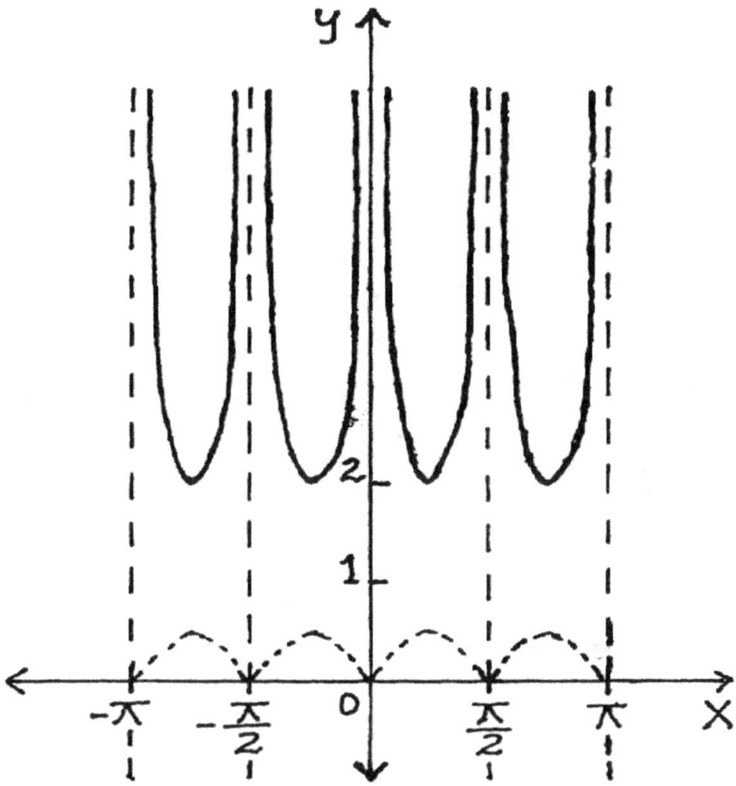

3. |sin|x|sinx|

Definition: Let x be a real number. Then the function h defined by

$$h(x) = |\sin|x|\sin x|$$

is called the absolute value of the product of the sine function and the sine of the absolute value function.

The domain of h is $(-\infty, \infty)$, and its range is [0, 1].

The domain of $\sin|x|\sin x$ is $(-\infty, \infty)$, and its range is [-1, 1].

EXAMPLE: $h(0) = 0$, $h(\pi/4) = 1/2$, $h(\pi/2) = 1$, $h(3\pi/4) = 1/2$, $h(\pi) = 0$, $h(5\pi/4) = 1/2$, $h(3\pi/2) = 1$, $h(7\pi/4) = 1/2$, $h(2\pi) = 0$, $h(-\pi/4) = 1/2$, $h(-\pi/2) = 1$, $h(-3\pi/4) = 1/2$, $h(-\pi) = 0$, $h(-5\pi/4) = 1/2$, $h(-3\pi/2) = 1$, $h(-7\pi/4) = 1/2$, and $h(-2\pi) = 0$. The graph of $y = h(x)$ is shown in Figure 8.3:

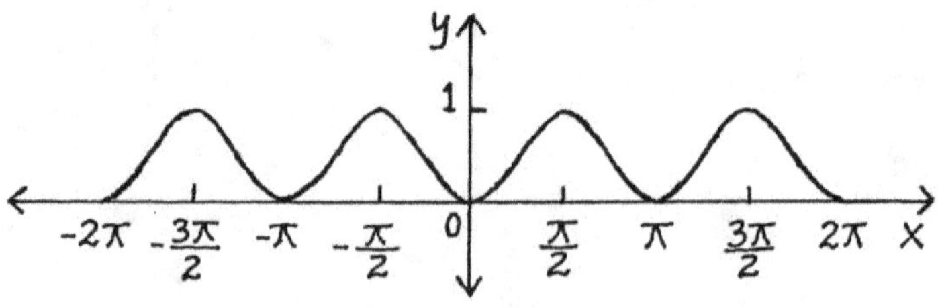

4. |csc|x|cscx|

Definition: Let x be a real number. Then the function h defined by

$$h(x) = |\csc|x|\csc|, \ \csc|x|\csc x = 1/(\sin|x|\sin x), \text{ and } \sin|x|\sin x \neq 0$$

is called the absolute value of the product of the cosecant function and the cosecant of the absolute value function.

The domain of h is {all real x: $x \neq \pm n\pi$, n = 0, 1, 2 . . .}, and the range of h is $[1, \infty)$. The vertical asymptotes of h are at $x = \pm n\pi$, n = 0, 1, 2 . . .

EXAMPLE: $h(\pi/4) = 2$, $h(\pi/2) = 1$, $h(3\pi/4) = 2$, $h(5\pi/4) = 2$, $h(3\pi/2) = 1$, $h(7\pi/4) = 2$, $h(-\pi/4) = 2$, $h(-\pi/2) = 1$, $h(-3\pi/4) = 2$, $h(-5\pi/4) = 2$, $h(-3\pi/2) = 1$, and $h(-7\pi/4) = 2$. The graphs of y = h(x), and y = |sin|x|sinx|, are shown in Figure 8.4:

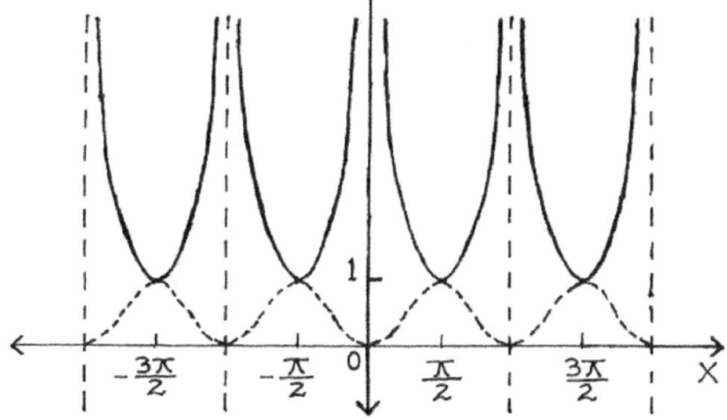

CHAPTER NINE

GRAPHS OF COSX/ | SINX | , SINX/ | COSX | , SIN | X | / | COSX | , | SINX | /COSX, | COSX | /SINX, AND | COSX | /SIN | X |

1. cosx/ | sinx |

Definition: Let x be a real number. Then the function h defined by

$$h(x) = \cos x/|\sin x|, \text{ for } |\sin x| \neq 0$$

is called the quotient of the cosine function and the absolute value of the sine function.

The domain of h is {all real x: x ≠ ± nπ, n = 0, 1, 2 . . .}, and the range of h is (-∞, ∞). The vertical asymptotes of h are at x = ± nπ, n = 0, 1, 2 . . .

EXAMPLE: h(π/4) = 1, h(π/2) = 0, h(3π/4) = -1, h(5π/4) = -1, h(3π/2) = 0, h(-π/4) = 1, h(-π/2) = 0, h(-3π/4) = -1, h(-5π/4) = -1, and h(-3π/2) = 0. The graph of y = h(x) is shown in Figure 9.1:

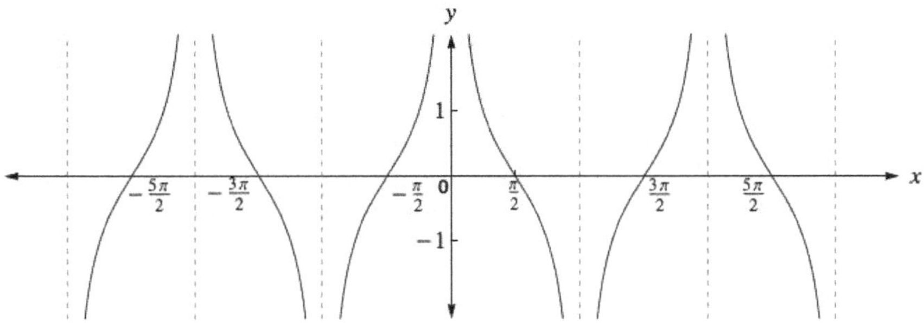

2. sinx/ | cosx |

Definition: Let x be a real number. Then the function h defined by

$$h(x) = \sin x/|\cos x|, \text{ for } |\cos x| \neq 0$$

is called the quotient of the sine function and the absolute value of the cosine function.

The domain of h is {all real x: x ≠ ± ($\frac{\pi}{2}$ + πn), n = 0, 1, 2 . . . }, and the range of h is (-∞, ∞). The vertical asymptotes are at x = ± ($\frac{\pi}{2}$ + πn), n = 0, 1, 2 . . .

EXAMPLE: h(0) = 0 h(π/4) = 1, h(3π/4) = 1, h(π) = 0, h(5π/4) = -1, h(-π/4) = -1, h(-3π/4) = -1, h(-π) = 0, and h(-5π/4) = 1. The graph of y = h(x) is shown in Figure 9.2:

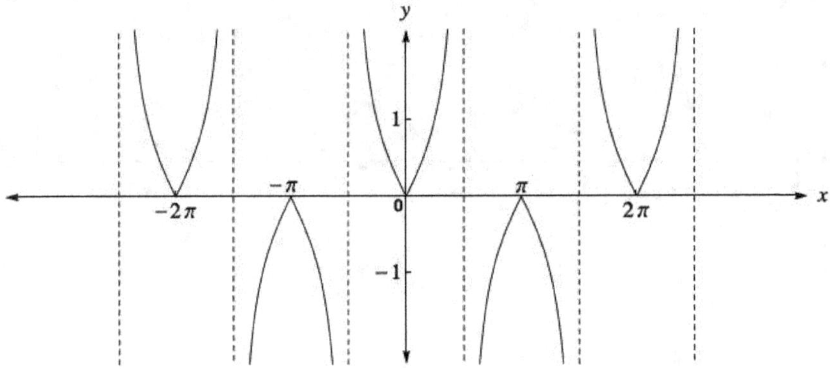

3. sin|x|/|cosx|

Definition: Let x be a real number. Then the function h defined by

$$h(x) = \sin|x|/|\cos x|, \text{ for } |\cos x| \neq 0$$

is called the quotient of the sine of the absolute value function and the absolute value of the cosine function.

The domain of h is {all real x: $x \neq \pm (\frac{\pi}{2} + \pi n)$, n = 0, 1, 2 . . . }, and the range of h is (-∞, ∞). The vertical asymptotes of h are at x = $\pm (\frac{\pi}{2} + \pi n)$, n = 0, 1, 2 . . .

EXAMPLE: h(0) = 0, h(π/4) = 1, h(3π/4) = 1, h(π) = 0, h(5π/4) = -1, h(-π/4) = 1, h(-3π/4) =1, h(-π) = 0 and h(-5π/4) = -1. The graph of y = h(x) is shown in Figure 9.3:

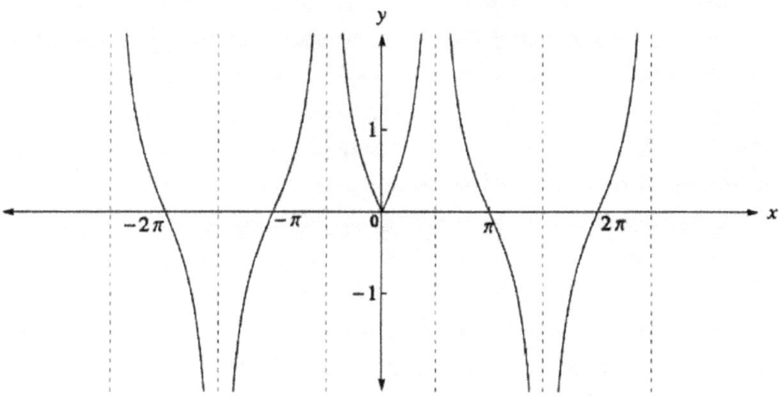

4. |sinx|/cosx

Definition: Let x be a real number. Then the function h defined by

$$h(x) = |\sin x|/\cos x, \text{ for } \cos x \neq 0$$

is called the quotient of the absolute value of the sine function and the cosine function.

The domain of h is {all real x: $x \neq \frac{\pi}{2} \pm n\pi$, n = 0, 1, 2 . . .}, and the range of h is $(-\infty, \infty)$. The vertical asymptotes of h are at $x = \frac{\pi}{2} \pm n\pi$, n = 0, 1, 2 . . .

EXAMPLE: h(0) = 0, h(π/4) = 1, h(3π/4) = -1, h(π) = 0, h(5π/4) = -1, h(7π/4) = 1, h(2π) = 0, h(9π/4) = 1, h(-π/4) = 1, h(-3π/4) = -1, h(-π) = 0, h(-5π/4) = -1, h(-7π/4) = 1, h(-2π) = 0, and h(-9π/4) = 1. The graph of y = h(x) is shown in Figure 9.4:

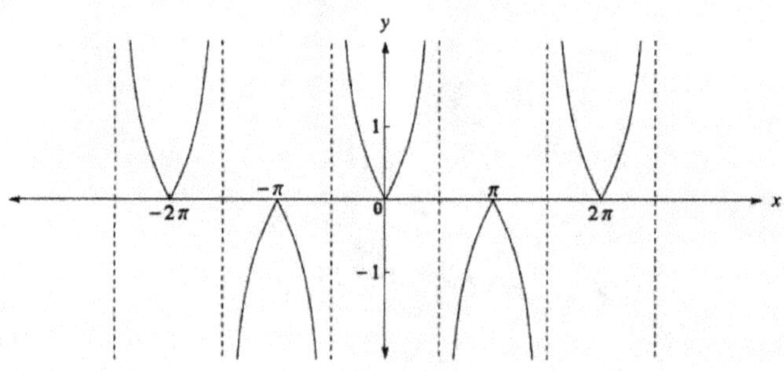

5. |cosx|/sinx

Definition: Let x be a real number. Then the function h defined by

$$h(x) = |\cos x|/\sin x, \text{ for } \sin x \neq 0$$

is called the quotient of the absolute value of the cosine function and the sine function.

The domain of h is {all real x: x $\neq \pm n\pi$, n = 0, 1, 2 . . .}, and the range of h is $(-\infty, \infty)$. The vertical asymptotes of h are at x = $\pm n\pi$, n = 0, 1, 2 . . .

EXAMPLE: h(π/4) = 1, h(π/2) = 0, h(3π/4) = 1, h(5π/4) = -1, h(3π/2) = 0, h(7π/4) = -1, h(9π/4) = 1, h(5π/2) = 0, h(-π/4) = -1, h(-π/2) = 0, h(-3π/4) = -1, h(-5π/4) = 1, h(-3π/2) = 0, h(-7π/4) = 1, h(-9π/4) = -1, and h(-5π/2) = 0. The graph of y = h(x) is shown in Figure 9.5:

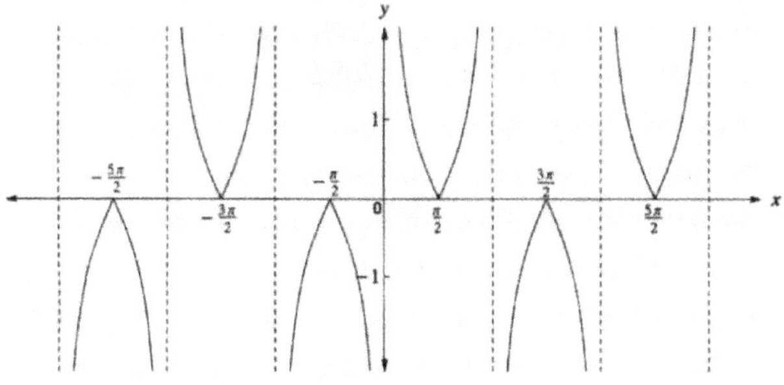

6. |cosx|/sin|x|

Definition: Let x be a real number. Then the function h defined by

$$h(x) = |\cos x|/\sin|x|, \text{ for } \sin|x| \neq 0$$

is called the quotient of the absolute value of the cosine function and the sine of the absolute value function.

The domain of h is {all real x: $x \neq \pm n\pi$, n = 0, 1, 2 . . .}, and the range of h is $(-\infty, \infty)$. The vertical asymptotes of h are at $x = \pm n\pi$, n = 0, 1, 2 . . .

EXAMPLE: $h(\pi/4) = 1$, $h(\pi/2) = 0$, $h(3\pi/4) = 1$, $h(5\pi/4) = -1$, $h(3\pi/2) = 0$, $h(7\pi/4) = -1$, $h(9\pi/4) = 1$, $h(5\pi/2) = 0$, $h(-\pi/4) = 1$, $h(-\pi/2) = 0$, $h(-3\pi/4) = 1$, $h(-5\pi/4) = -1$, $h(-3\pi/2) = 0$, $h(-7\pi/4) = -1$, $h(-9\pi/4) = 1$, and $h(-5\pi/2) = 0$. The graph of $y = h(x)$ is shown in Figure 9.6:

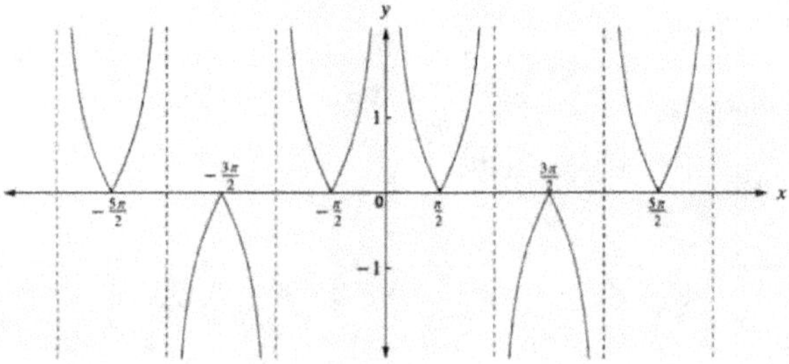

CHAPTER TEN

GRAPHS OF TANX/COSX, TANX/|COSX|, |TANX|/COSX, TAN|X|/COSX, AND TAN|X|/|COSX|

1. tanx/cosx

Definition: Let x be a real number. Then the function h defined by

$$h(x) = \tan x/\cos x \text{ and } \cos x \neq 0$$

is called the quotient of the tangent function and the cosine function.

The domain of h is {all real x: $x \neq \frac{\pi}{2} \pm n\pi$, n = 0, 1, 2,...}, and the range of h is $(-\infty, \infty)$. The vertical asymptotes of h are at $x = \frac{\pi}{2} \pm n\pi$, n = 0, 1, 2,...

EXAMPLE: h(0) = 0, h(π/6) = 2/3, h(5π/6) = 2/3, h(π) = 0, h(7π/6) = -2/3, h(11π/6) = -2/3, h(2π) = 0, h(-π/6) = -2/3, h(-5π/6) = -2/3, h(-π) = 0, h(-7π/6) = 2/3, h(-11π/6) = 2/3, and h(-2π) = 0. The graph of y = h(x) is shown in Figure 10.1:

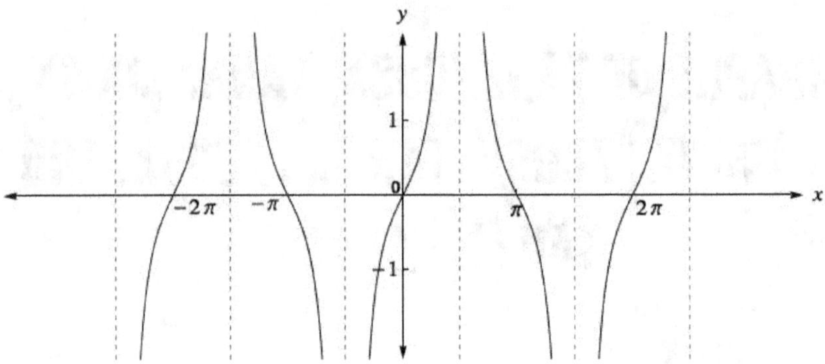

2. tanx/|cosx|

Definition: Let x be a real number. Then the function h defined by

$$h(x) = \tan x/|\cos x| \text{ and } |\cos x| \neq 0$$

is called the quotient of the tangent function and the absolute value of the cosine function.

The domain of h is {all real x: x $\neq \pm (\frac{\pi}{2} + \pi n)$, n = 0, 1, 2 . . .), n = 0, 1, 2 . . .}, and the range of h is (-∞, ∞). The vertical asymptotes of h are at x = $\pm (\frac{\pi}{2} + \pi n)$, n = 0, 1, 2 . . .

EXAMPLE: h(0) = 0, h(π/6) = 2/3, h(π/4) = $\sqrt{2}$, h(5π/6) = -2/3, h(π) = 0, h(7π/6) = 2/3, h(-π/6) = -2/3, h(-π/4) = -$\sqrt{2}$, h(-5π/6) = 2/3, h(-π) = 0, and h(-7π/) = -2/3. The graph of y = h(x) is shown in Figure 10.2:

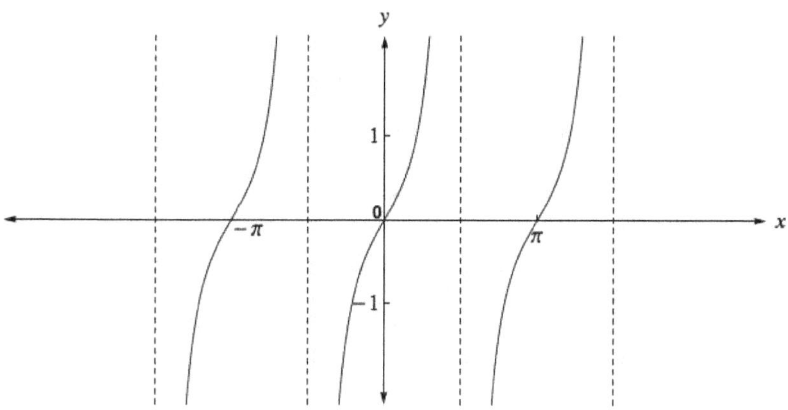

3. |tanx| /cosx

Definition: Let x be a real number. Then the function h defined by

$$h(x) = |\tan x|/\cos x, \text{ and } \cos x \neq 0$$

is called the quotient of the absolute value of the tangent function and the cosine function.

The domain of h is {all real x: $x \neq \pm(\frac{\pi}{2} \pm n\pi)$, n = 0, 1, 2 . . .}, and the range of h is $(-\infty, \infty)$. The vertical asymptotes of h are at $x = \frac{\pi}{2} \pm n\pi$, n = 0, 1, 2 . . .

EXAMPLE: h(0) = 0, h(π/6) = 2/3, h(5π/6) = -2/3, h(π) = 0, h(7π/6) = -2/3, h(11π/6) = 2/3, h(2π) = 0, h(-π/6) = 2/3, h(-5π/6) = -2/3, h(-π) =0, h(-7π/6) = -2/3, h(-11π/6) = 2/3, and h(-2π) = 0. The graph of y = h(x) is shown in Figure 10.3:

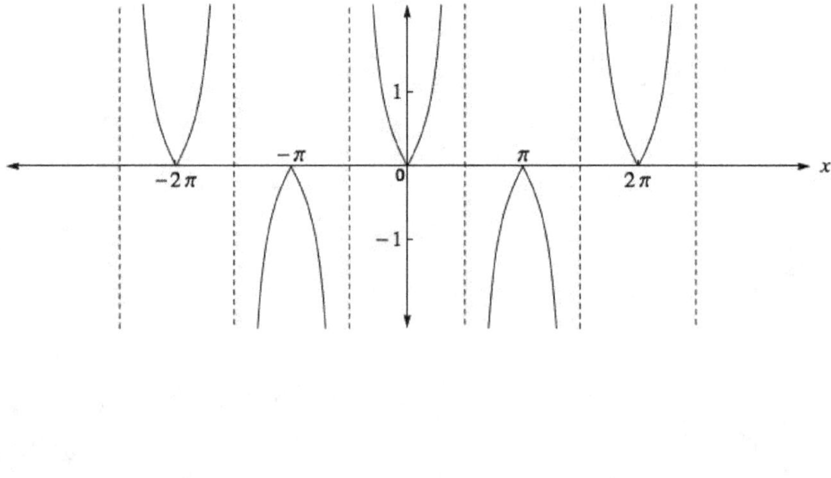

4. tan|x|/cosx

Definition: Let x be a real number. Then the function h defined by

$$h(x) = \tan|x|/\cos x \text{ and } \cos x \neq 0$$

is called the quotient of the tangent of the absolute value function and the cosine function.

The domain of h is {all real x: $x \neq \frac{\pi}{2} \pm n\pi$, n = 0, 1, 2 . . .}, and the range of h is $(-\infty, \infty)$. The vertical asymptotes of h are at $x = \frac{\pi}{2} \pm n\pi$, n = 0, 1, 2 . . .

EXAMPLE: h(0) = 0, h(π/6) = 2/3, h(5π/6) = 2/3, h(π) = 0, h(7π/6) = -2/3, h(11π/6) = -2/3, h(2π) = 0, h(-π/6) = 2/3, h(-5π/6) = 2/3, h(-π) = 0, h(-7π/6) = -2/3, h(-11π/6) = -2/3, and h(-2π) = 0. The graph of y = h(x) is shown in Figure 10.4:

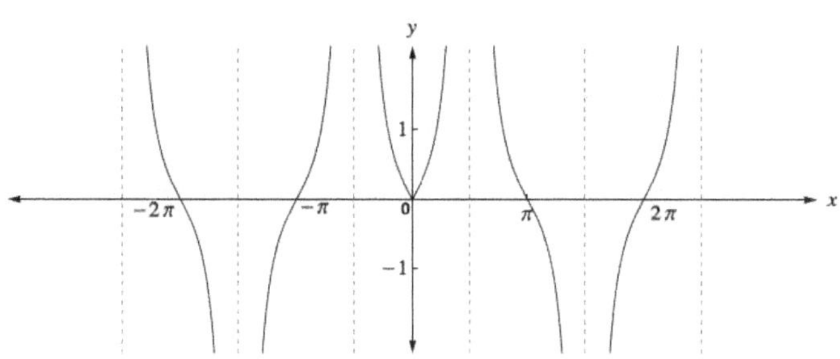

5. tan|x|/|cosx|

Definition: Let x be a real number. Then the function h defined by

$$h(x) = \tan|x|/|\cos x|, \text{ and } \cos x \neq 0$$

is called the quotient of the tangent of the absolute value function and the absolute value of the cosine function.

The domain of h is {all real x: $x \neq \frac{\pi}{2} \pm n\pi$, n = 0, 1, 2 . . .}, and the range of h is $(-\infty, \infty)$. The vertical asymptotes of h are at $x = \frac{\pi}{2} \pm n\pi$, n = 0, 1, 2 . . .

EXAMPLE: h(0) = 0, h(π/6) = 2/3, h(5π/6) = -2/3, h(π) = 0, h(7π/6) = 2/3, h(11π/6) = -2/3, h(2π) = 0, h(-π/6) = 2/3, h(-5π/6) = -2/3, h(-π) = 0, h(-7π/6) = 2/3, h(-11π/6) = -2/3, and h(-2π) = 0. The graph of y = h(x) is shown in Figure 10.5:

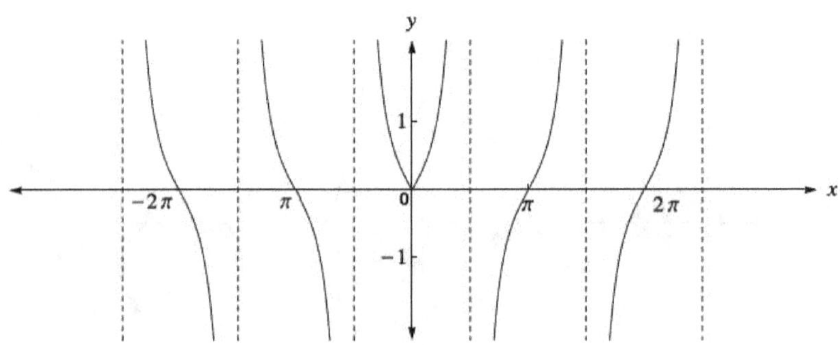

CHAPTER ELEVEN

GRAPHS OF SINX + COSX, SINX - COSX, COSX - SINX, SIN|X| + COSX, SIN|X| - COSX, AND COSX - SIN|X|

1. sinx + cosx

Definition: Let x be a real number. Then the function h defined by

$$h(x) = \sin x + \cos x$$

is called the sum of the sine and cosine functions.

The domain of h is $(-\infty, \infty)$, and its range is $[-\sqrt{2}, \sqrt{2}\,]$.

EXAMPLE: $h(0) = 1$, $h(\pi/2) = 1$, $h(\pi) = -1$, $h(3\pi/2) = -1$, $h(2\pi) = 1$, $h(-\pi/2) = -1$, $h(-\pi) = -1$, $h(-3\pi/2) = 1$, and $h(-2\pi) = 1$. The graph of $y = h(x)$ is shown in Figure 11.1:

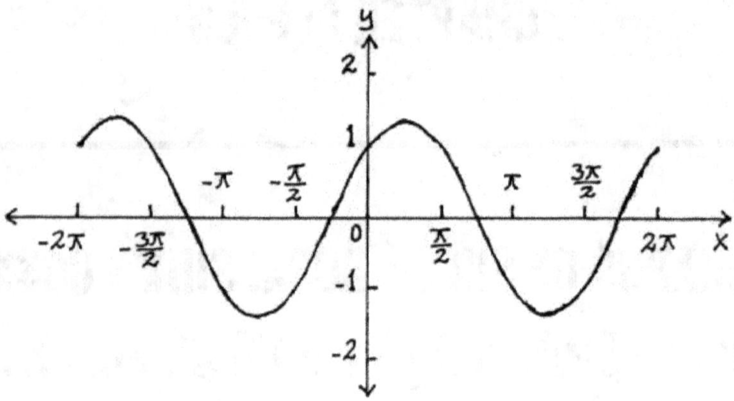

2. sinx - cosx

Definition: Let x be a real number. Then the function h defined by

$$h(x) = \sin x - \cos x$$

is called the difference of the sine and cosine functions.

The domain of h is $(-\infty, \infty)$, and its range is $[-\sqrt{2}, \sqrt{2}\,]$.

EXAMPLE: $h(0) = -1$, $h(\pi/2) = 1$, $h(\pi) = 1$, $h(3\pi/2) = -1$, $h(2\pi) = -1$, $h(-\pi/2) = -1$, $h(-\pi) = 1$, $h(-3\pi/2) = 1$, and $h(-2\pi) = -1$. The graph of $y = h(x)$ is shown in Figure 11.2:

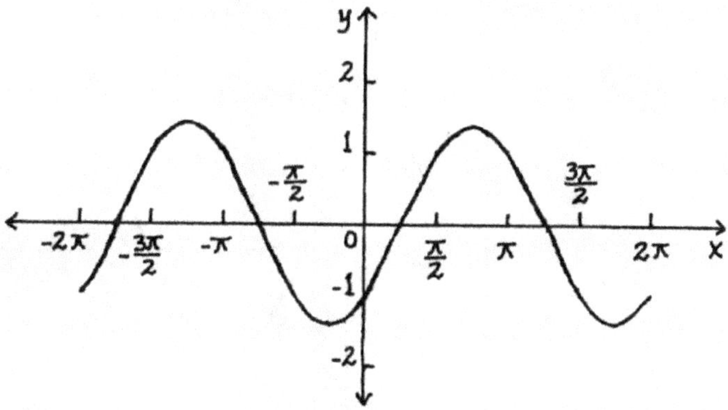

3. cosx – sinx

Definition: Let x be a real number. Then the function h defined by

$$h(x) = \cos x - \sin x$$

is called the difference of the cosine and sine functions.

The domain of h is $(-\infty, \infty)$, and its range is $[-\sqrt{2}, \sqrt{2}\,]$.

EXAMPLE: $h(0) = 1$, $h(\pi/2) = -1$, $h(\pi) = -1$, $h(3\pi/2) = 1$, $h(2\pi) = 1$, $h(-\pi/2) = 1$, $h(-\pi) = -1$, $h(-3\pi/2) = -1$, and $h(-2\pi) = 1$. The graph of $y = h(x)$ is shown in Figure 11.3:

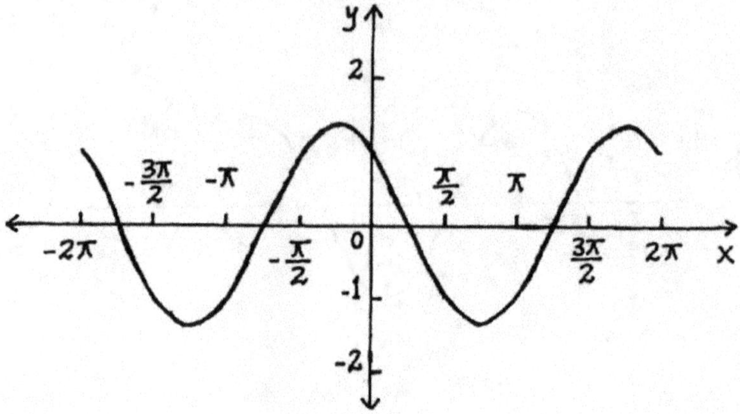

4. sin|x| + cosx

Definition: Let x be a real number. Then the function h defined by

$$h(x) = \sin|x| + \cos x$$

is called the sum of the sine of the absolute value function and the cosine function.

The domain of h is $(-\infty, \infty)$, and its range is $[-\sqrt{2}, \sqrt{2}\,]$.

EXAMPLE: h(0) = 1, h(π/2) = 1, h(π) = -1, h(3π/2) = -1, h(2π) = 1, h(-π/2) = 1, h(-π) = -1, h(-3π/2) = -1, and h(-2π) = 1. The graph of y = h(x) is shown in Figure 11.4:

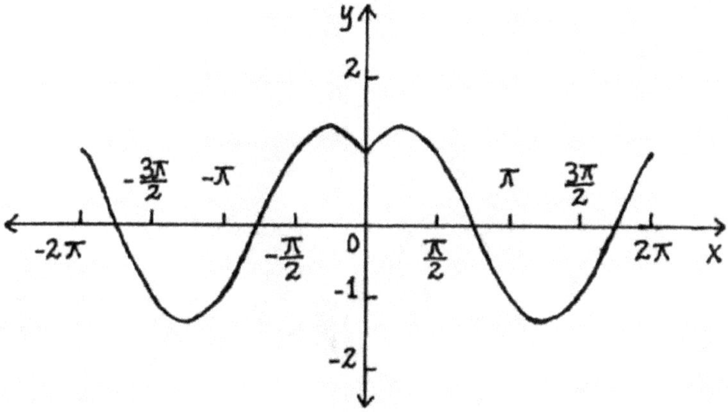

5. sin|x| - cosx

Definition: Let x be a real number. Then the function h defined by

$$h(x) = \sin|x| - \cos x$$

is called the difference of the sine of the absolute value function and the cosine function.

The domain of h is $(-\infty, \infty)$, and its range is $[-\sqrt{2}, \sqrt{2}\,]$.

EXAMPLE: $h(0) = -1$, $h(\pi/2) = 1$, $h(\pi) = 1$, $h(3\pi/2) = -1$, $h(2\pi) = -1$, $h(-\pi/2) = 1$, $h(-\pi) = 1$, $h(-3\pi/2) = -1$, and $h(-2\pi) = -1$. The graph of $y = h(x)$ is shown in Figure 11.5:

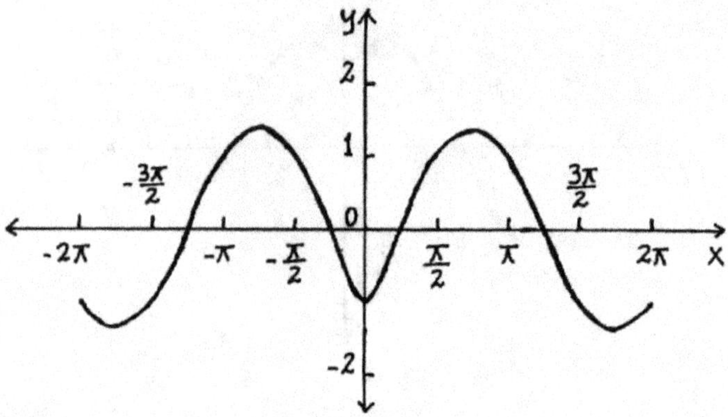

6. cosx - sin|x|

Definition: Let x be a real number. Then the function h defined by

$$h(x) = \cos x - \sin|x|$$

is called the difference of the cosine function and the sine of the absolute value function.

The domain of h is $(-\infty, \infty)$, and its range is $[-\sqrt{2}, \sqrt{2}\,]$.

EXAMPLE: $h(0) = 1$, $h(\pi/2) = -1$, $h(\pi) = -1$, $h(3\pi/2) = 1$, $h(2\pi) = 1$, $h(-\pi/2) = -1$, $h(-\pi) = -1$, $h(3\pi/2) = 1$, and $h(-2\pi) = 1$. The graph of $y = h(x)$ is shown in Figure 11.6:

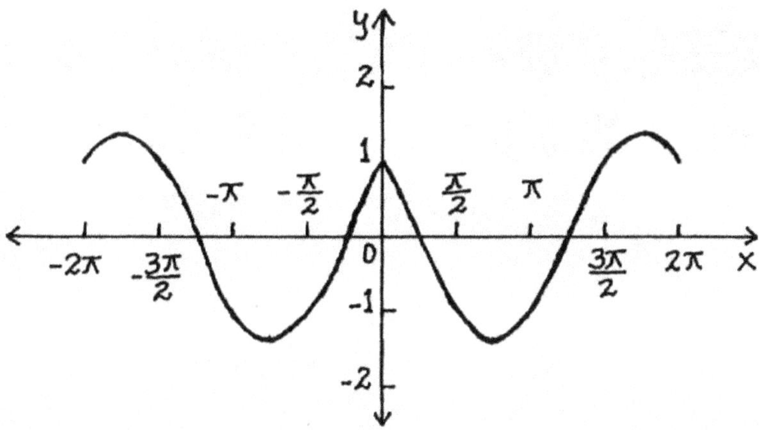

CHAPTER TWELVE

GRAPHS OF (SINX - COSX)/(SINX + COSX), (COSX - SINX)/(SINX + COSX), (SINX + COSX)/(SINX - COSX), AND (SINX + COSX)/(COSX - SINX)

1. (sinx - cosx)/(sinx + cosx)

Definition: Let x be a real number. Then the function h defined by

$$h(x) = (\sin x - \cos x)/(\sin x + \cos x), \text{ for } \sin x + \cos x \neq 0$$

is called the quotient of the difference of the sine and cosine functions and the sum of the sine and cosine functions.

The domain of h is {all real x: $x \neq -\frac{\pi}{4} \pm n\pi$, n = 0, 1, 2 . . .}, and the range of h is $(-\infty, \infty)$. The vertical asymptotes of h are at $x = -\frac{\pi}{4} \pm n\pi$, n = 0, 1, 2 . . . The x-intercepts of h are at $x = \frac{\pi}{4} \pm n\pi$, n = 0, 1, 2 . .

EXAMPLE: h(0) = -1, h(π/2) = 1, h(π) = -1, h(3π/2) = 1, h(2π) = -1, h(-π/2) = 1, h(-π) = -1, h(-3π/2) = 1, and h(-2π) = -1.The graph of y = h(x) is shown in Figure 12.1:

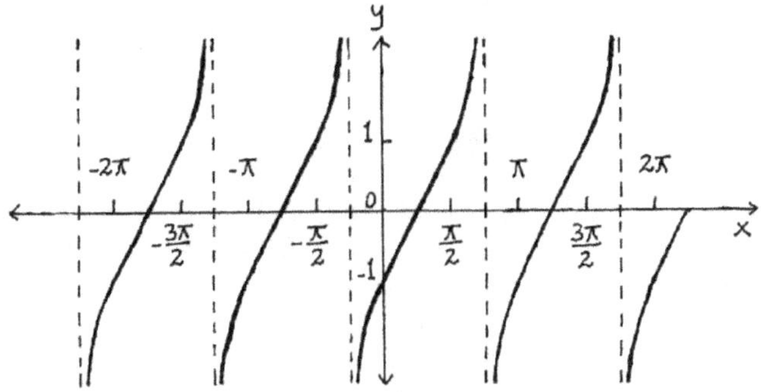

2. (cosx - sinx)/(sinx + cosx)

Definition: Let x be a real number. Then the function h defined by

h(x) = (cosx - sinx)/(sinx +cosx), for sinx + cosx ≠ 0

is called the quotient of the difference of the cosine and sine functions and the sum of the sine and cosine functions.

The domain of h is {all real x: x ≠ $-\frac{\pi}{4}$ ± nπ, n = 0, 1, 2 . . .}, and the range of h is (-∞, ∞). The vertical asymptotes of h are at x = $-\frac{\pi}{4}$ ± nπ, n = 0, 1, 2 . . . The x-intercepts of h are at x = $\frac{\pi}{4}$ ± nπ, n = 0, 1, 2 . . .

EXAMPLE: $h(0) = 1$, $h(\pi/2) = -1$, $h(\pi) = 1$, $h(3\pi/2) = -1$, $h(2\pi) = 1$, $h(-\pi/2) = -1$, $h(-\pi) = 1$, $h(-3\pi/2) = -1$, and $h(-2\pi) = 1$. The graph of $y = h(x)$ is shown in Figure 12.2:

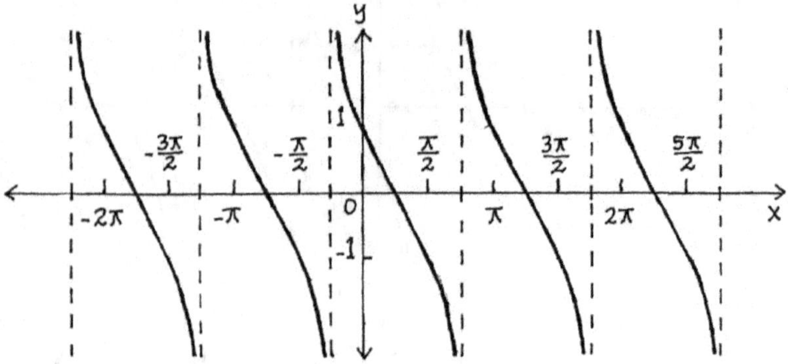

3. (sinx + cosx)/(sinx - cosx)

Definition: Let x be a real number. Then the function h defined by

$$h(x) = (\sin x + \cos x)/(\sin x - \cos x), \text{ for } \sin x - \cos x \neq 0$$

is called the quotient of the sum of the sine and cosine functions and the difference of the sine and cosine functions.

The domain of h is {all real x: $x \neq \frac{\pi}{4} \pm n\pi$, n = 0, 1, 2 . . .}, and the range of h is (-∞, ∞). The vertical asymptotes of h are at $x = \frac{\pi}{4} \pm n\pi$, n = 0, 1, 2 . . . The x-intercepts of h are at $x = -\frac{\pi}{4} \pm n\pi$, n = 0, 1, 2 . .

EXAMPLE: h(0) = -1, h(π/2) 1, h(π) = -1, h(3π/2) = 1, h(2π) = -1, h(-π/2) = 1, h(-π) = -1, h(-3π/2) = 1, and h(-2π) = -1. The graph of y = h(x) is shown in Figure 12.3:

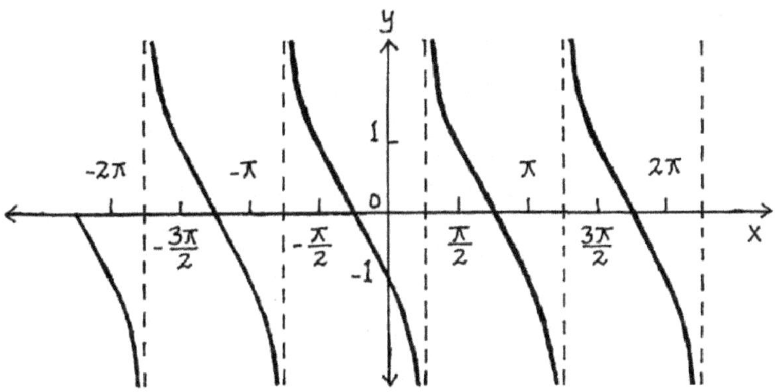

4. (sinx + cosx)/(cosx - sinx)

Definition: Let x be a real number. Then the function h defined by

h(x) = (sinx + cosx)/(cos - sinx), for sinx - cosx ≠ 0

is called the quotient of the sum of the sine and cosine functions and the difference of the cosine and sine functions.

The domain of h is {all real x: $x \neq \frac{\pi}{4} \pm n\pi$, n = 0, 1, 2 . . .}, and the range of h is $(-\infty, \infty)$. The vertical asymptotes of h are at $x = \frac{\pi}{4} \pm n\pi$, n = 0, 1, 2 . . . The x-intercepts of h are at $x = -\frac{\pi}{4} \pm n\pi$, n = 0, 1, 2 . . .

EXAMPLE: h(0) = 1, h(π/2) = -1, h(π) = 1, h(3π/2) = -1, h(2π) = 1, h(-π/2) = -1, h(-π) = 1, h(-3π/2) = -1, and h(-2π) = 1. The graph of y = h(x) is shown in Figure 12.4:

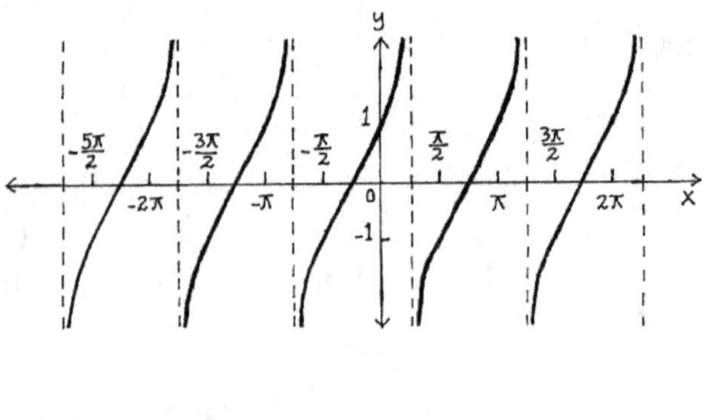

CHAPTER THIRTEEN

GRAPHS OF SINX/(SINX - COSX), SINX/(COSX - SINX), SINX/(SINX + COSX), COSX/(SINX - COSX), COSX/(COSX - SINX), AND COSX/(SINX + COSX)

1. sinx/(sinx - cosx)

Definition: Let x be a real number. Then the function h defined by

$$h(x) = \sin x/(\sin x - \cos x), \text{ for } \sin x - \cos x \neq 0$$

is called the quotient of the sine function and the difference of the sine and cosine functions.

The domain of h is {all real: $x \neq \frac{\pi}{4} \pm n\pi$, n = 0, 1, 2 . . .}, and the range of h is (-∞, ∞). The vertical asymptotes of h are at $x = \frac{\pi}{4} \pm n\pi$, n = 0, 1, 2 . . . The x-intercepts of h are at x = ± nπ, n = 0, 1, 2 . . .

EXAMPLE: h(0) = 0, h(π/2) = 1, h(3π/4) =1/2, h(π) = 0, h(3π/2) = 1, h(7π/4) = 1/2, h(2π) = 0, h(-π/4) = 1/2, h(-π/2) = 1, h(-π) = 0, h(-5π/4) = 1/2, h(-3π/2) = 1, and h(-2π) = 0. The graph of y = h(x) is shown in Figure 13.1:

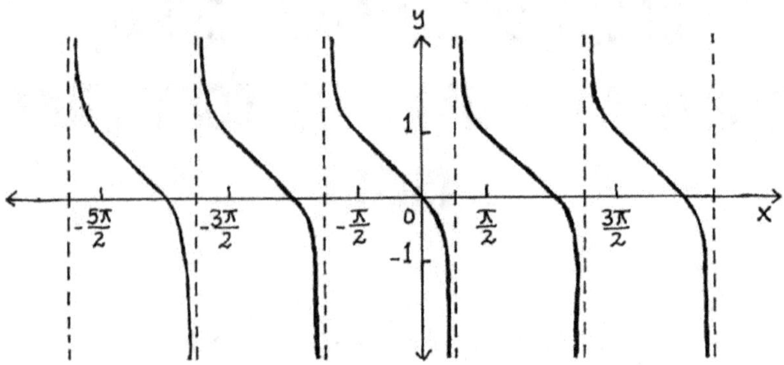

2. sinx/(cosx - sinx)

Definition: Let x be a real number. Then the function h defined by

$$h(x) = \sin x/(\cos x - \sin x), \text{ for } \cos x - \sin x \neq 0$$

is called the quotient of the sine function and the difference of the cosine and sine functions.

The domain of h is {all real x: x $\neq \frac{\pi}{4} \pm n\pi$, n = 0, 1, 2 . . .}, and the range of h is (-∞, ∞). The vertical asymptotes of h are at x = $\frac{\pi}{4}$ $\pm n\pi$, n = 0, 1, 2 . . . The x-intercepts of h are at x = $\pm n\pi$, n = 0, 1, 2 . . .

EXAMPLE: h(0) = 0, h(π/2) = -1, h(3π/4) = -1/2, h(π) = 0, h(3π/2) = -1, h(7π/4) = -1/2, h(2π) = 0, h(-π/4) = -1/2, h(-π/2) = -1, h(-π) = 0, h(-5π/4) = -1/2, h(-3π/2) = -1, and h(-2π) = 0. = -1. The graph of y = h(x) is shown in Figure 13.2:

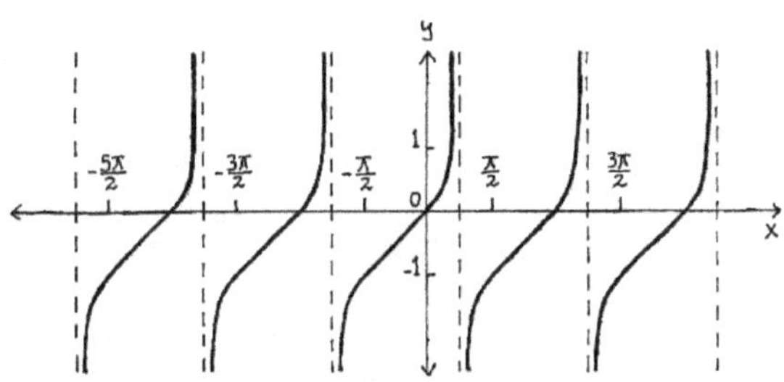

3. sinx/(sinx + cosx)

Definition: Let x be a real number. Then the function h defined by

$$h(x) = \sin x/(\sin x + \cos x), \text{ for } \sin x + \cos x \neq 0$$

is called the quotient of the sine function and the sum of the sine and cosine functions.

The domain of h is {all real x: $x \neq -\frac{\pi}{4} \pm n\pi$, n = 0, 1, 2 . . .}, and the range of h is $(-\infty, \infty)$. The vertical asymptotes of h are at $x = -\frac{\pi}{4} \pm n\pi$, n = 0, 1, 2 . . . The x-intercepts of h are at $x = \pm n\pi$, n = 0, 1, 2 . . .

EXAMPLE: h(0) = 0, h(π/4) = 1/2, h(π/2) = 1, h(π) = 0, h(5π/4) = 1/2, h(3π/2) = 1, h(2π) = 0, h(-π/2) = 1, h(-3π/4) = 1/2, h(-π) = 0, h(-3π/2) = 1, h(-7π/4) = 1/2, and h(-2π) = 0. The graph of y = h(x) is shown in Figure 13.3:

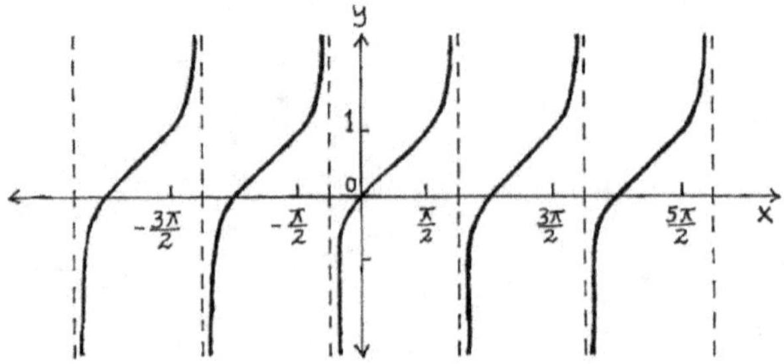

4. cosx/(sinx - cosx)

Definition: Let x be a real number. Then the function h defined by

$$h(x) = \cos x/(\sin x - \cos x), \text{ for } \sin x - \cos x \neq 0$$

is called the quotient of the cosine function and the difference of the sine and cosine functions.

The domain of h is {all real x: $x \neq \frac{\pi}{4} \pm n\pi$, n = 0, 1, 2 . . .}, and the range of h is $(-\infty, \infty)$. The vertical asymptotes of h are at $x = \frac{\pi}{4} \pm n\pi$, n = 0, 1, 2 . . . The x-intercepts of h are at $x = \frac{\pi}{2} \pm n\pi$, n = 0, 1, 2 . . .

EXAMPLE: h(0) = -1, h(π/2) = 0, h(3π/4) = -1/2, h(π) = -1, h(3π/2) = 0, h(7π/4) = -1/2, h(2π) = -1, h(-π/4) = -1/2, h(-π/2) = 0, h(-π) = -1, h(-5π/4) = -1/2, h(-3π/2) = 0, and h(-2π) = -1. The graph of y = h(x) is shown in Figure 13.4:

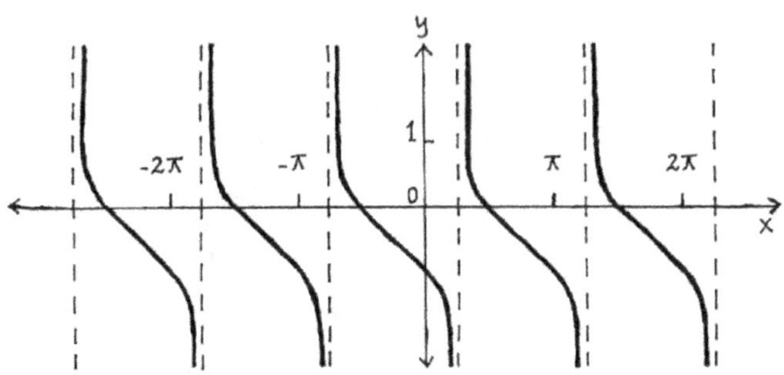

5. cosx/(cosx – sinx)

Definition: Let x be a real number. Then the function h defined by

$$h(x) = \cos x/(\cos x - \sin x), \text{ for } \cos x - \sin x \neq 0$$

is called the quotient of the cosine function and the difference of the cosine and sine functions.

The domain of h is {all real x : $x \neq \frac{\pi}{4} \pm n\pi$, n = 0, 1, 2 . . .}, and the range of h is $(-\infty, \infty)$. The vertical asymptotes of h are at $x = \frac{\pi}{4} \pm n\pi$, n = 0, 1, 2 . . . The x-intercepts of h are at $x = \frac{\pi}{2} \pm n\pi$, n = 0, 1, 2 . . .

EXAMPLE: h(0) = 1, h(π/2) = 0, h(3π/4) = 1/2, h(π) = 1, h(3π/2) = 0, h(7π/4) = 1/2, h(2π) = 1, h(-π/4) = 1/2, h(-π/2) = 0, h(-π) = 1, h(-5π/4) = 1/2, h(-3π/2) = 0, and h(-2π) = 1. The graph of y = h(x) is shown in Figure 13.5:

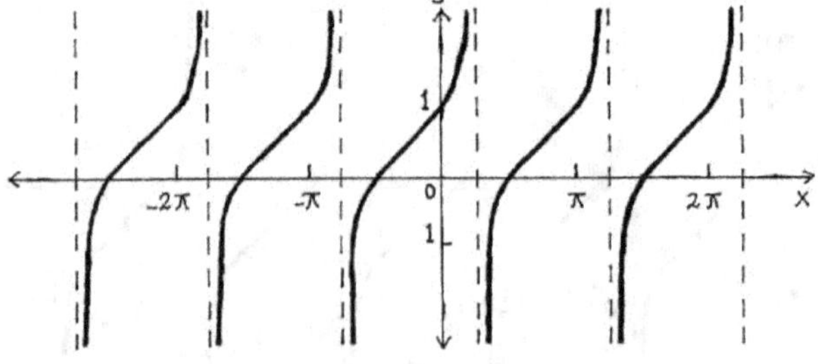

6. cosx/(sinx + cosx)

Definition: Let x be a real number. Then the function h defined by

$$h(x) = \cos x/(\sin x + \cos x), \text{ for } \sin x + \cos x \neq 0$$

is called the quotient of the cosine function and the sum of the sine and cosine functions.

The domain of h is {all real x: $x \neq -\frac{\pi}{4} \pm n\pi$, n = 0, 1, 2 . . .}, and the range of h is $(-\infty, \infty)$. The vertical asymptotes of h are at $x = -\frac{\pi}{4} \pm n\pi$, n = 0, 1, 2 . . . The x-intercepts of h are at $x = \frac{\pi}{2} \pm n\pi$, n = 0, 1, 2 . . .

EXAMPLE: h(0) = 1, h(π/4) = 1/2, h(π/2) = 0, h(π) = 1, h(5π/4) = 1/2, h(3π/2) = 0, h(2π) = 1, h(-π/2) = 0, h(-3π/4) = 1/2, h(-π) = 1, h(-3π/2) = 0, h(-7π/4) = 1/2, and h(-2π) = 1. The graph of y = h(x) is shown in Figure 13.6:

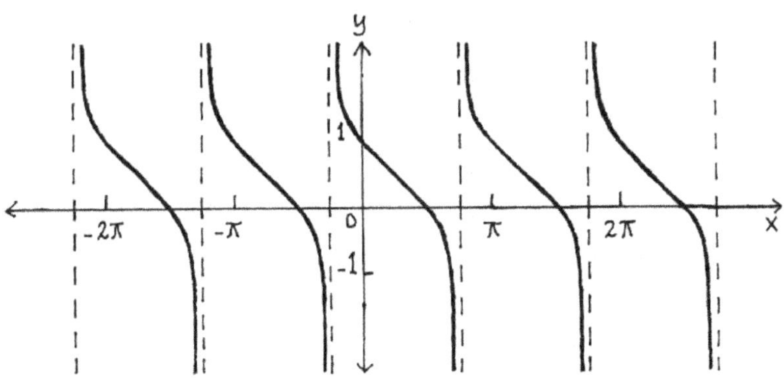

RELATED BOOKS

Downing, D. E-Z Trigonometry, Barron's Educational Series, Inc., 250 Wireless Blvd., Hauppauge, New York, 2010.

Gibilisco, S. Trigonometry Demystified, Second Edition, The McGraw-Hill Companies, Inc., New York, NY, 2012.

Huettenmueller, R. Pre-Calculus Demystified, Second Edition, The McGraw-Hill Companies, Inc., New York, NY, 2012.

Krantz, S.G. Calculus Demystified, Second Edition, The McGraw-Hill Companies, Inc., New York, NY, 2011.

Kuang, Y., and E. Kase. Pre-Calculus For Dummies, Second Edition, John Wiley & Sons, Inc., 111 River Street., Hoboken, NJ 07030-5774, 2012.

Leff, S.L. E-Z Precalculus, Barron's Educational Series, Inc., 250 Wireless Blvd., Hauppauge, New York, 2010.

Moyer, E.R., and F. Ayers. Schaum's Outline of Trigonometry, Fourth Edition, The McGraw-Hill Companies, Inc., New York, NY, 2009.

Safier, F. Schaum's Outline of Precalculus, Second Edition, The McGraw-Hill Companies, Inc., New York, NY, 2009.

Sterling, M.J. Trigonometry For Dummies, 3rd Edition, John Wiley & Sons, Inc., 111 River Street, Hoboken, NJ 07030-5774, 2023.

Sterling, M.J. Trigonometry For Dummies, Second Edition, John Wiley & Sons, Inc., 111 River St., Hoboken, NJ 07030-5774, 2014.

Sterling, M.J. Precalculus For Dummies, 3rd Edition, John Wiley & Sons, Inc., 111 River Street, Hoboken, NJ 07030-5774, 2019.

Sterling, MJ. Precalculus: !001 Practice Problems For Dummies, John Wiley & Sons, Inc., 111 River Street, Hoboken, NJ 07030-5774, 2022.

ABOUT THE BOOK

This book is Volume 1 of new original graphs invented in trigonometry by Chibamba Mulenga, Ph.D. They are graphs for which the author received copyrights from the United States Library of Congress Copyright Office. Through prompting by his inner voice, the author first invented the function $\sin|x|$ and its graph, where the x is a real number and the expression $|x|$ is the absolute value of x. Inspired by inventing $\sin|x|$ and its graph, the author subsequently formulated other suitable trigonometric functions and their graphs. Through its concepts, definitions, examples, and graphs, the book has great potential to inspire its readers to pursue careers in fields such as mathematics, science, engineering, astronomy, land surveying, oceanography, and weather forecasting. The book is suitable for students who have studied trigonometry. The book can serve as a pedagogical resource in high school and college mathematics curricula. Mathematicians can use the book as a starting point from which to open up new lines of research in trigonometry.

ABOUT THE AUTHOR

Dr. Chibamba Mulenga is a distinguished mathematician, educator, and inventor. He is the author of The Fundamental Principle of Digits of a Number and has made significant contributions to the field of mathematics education. Dr. Mulenga earned his Ph.D. in Mathematics Education from The Ohio State University, where he also served as a Graduate Teaching Associate in the Department of Mathematics. He holds a Bachelor of Science in Computer Science from the same institution and a Master of Science in Mathematics and Statistics from Case Western Reserve University in Cleveland, Ohio. An innovative thinker, Dr. Mulenga has created over 130 original trigonometric graphs, all of which are registered with the United States Library of Congress. He is also the inventor of the Smallest Integer Function, for which he holds a U.S. copyright. His work reflects a deep commitment to advancing mathematical understanding through original research, creative problem-solving and groundbreaking discoveries.

www.ingramcontent.com/pod-product-compliance
Lightning Source LLC
LaVergne TN
LVHW022013200926
848481LV00007B/2365